SOME OF WHAT YOU'RE ABOUT TO LEARN MAY SURPRISE YOU. . . .

DISCOVER OVER 100 TIPS TO KEEP YOUR FAMILY SAFE, INCLUDING:

★ **Wash bananas but *do not* wash raw chicken. . . .** Your fingers or a knife can transport contaminants from the banana peel to the edible interior. Rinsing raw meat, chicken, or seafood cannot remove the bacteria—only thorough cooking can—but it *can* spread the bacteria around your sink!

★ **Do not store highly perishable foods—including eggs and milk—in the refrigerator door. . . .** It's true. Door temperatures can be slightly higher than inside the refrigerator's main compartment.

★ **Keep yourself and your food at least six feet away from the electronic bug zapper when you're eating in your backyard. . . .** Millions of bacteria are scattered into the air when a common housefly gets zapped!

★ **Do not use a fork to turn chickens, steaks, or other solid meats when they are cooking. . . .** Bacteria are generally confined to the surface of solid cuts of meat. By piercing the meat with a fork, you can transfer surface contamination to the sterile interior.

★ **PLUS: Sanitize your sponge by microwaving it on high for one minute. . . . Bag cold items together to provide a measure of insulation. . . . Use liquid soap instead of bar soap to prevent the spread of bacteria . . . and much more!**

GET SMART. GET THE FACTS, WITH

SAFE EATING

SAFE EATING

Protect Yourself Against *E. coli*, *Salmonella*, and Other Deadly Food-borne Pathogens

David W. K. Acheson, M.D.,
and Robin K. Levinson

A DELL BOOK

Published by
Dell Publishing
a division of
Bantam Doubleday Dell Publishing Group, Inc.
1540 Broadway
New York, New York 10036

ISBN: 0-440-22659-7

Printed in the United States of America

Published simultaneously in Canada

December 1998

10 9 8 7 6 5 4 3 2 1

OPM

For Genevieve,

who died at the age of four of a bacterial infection

believed to have been E. coli *O157:H7.*

Your mother, father, grandparents,

aunts, and uncles will

always love you.

And for Genevieve's pals—children everywhere.

May you be safe, healthy, and well loved.

CONTENTS

PART 1

Medical Aspects of Food-borne Diseases

PART 2

Keeping the Food Groups Safe

PART 3
Food-Safety Guidelines

ACKNOWLEDGMENTS

Many generous, community-spirited people helped make this book a reality. Our heartfelt thanks go to Angela Dansby, media relations specialist, Institute of Food Technologists, for all her enthusiastic assistance.

The following food-safety experts reviewed the manuscript and offered invaluable feedback:

Robert E. Anderson, Ph.D., professor emeritus, Division of Plant Sciences, West Virginia University

David Battigelli, Ph.D., adjunct professor of preventive medicine, University of Wisconsin at Madison, and chief of environmental virology, Wisconsin State Laboratory of Hygiene

Christine M. Bruhn, Ph.D., director, Center for Consumer Research, University of California at Davis

Dean O. Cliver, Ph.D., professor, Department of Population Health and Reproduction, School of Veterinary Medicine, University of California at Davis

Daniel Y.C. Fung, Ph.D., professor, Department of Animal Science, Kansas State University

Laura Garrido, M.S., seafood nutrition specialist, Bureau of Seafood and Aquaculture, University of Florida

Robert M. Grodner, Ph.D., professor, Department of Food Science, Louisiana State University

Linda J. Harris, Ph.D., extension specialist in microbial food safety, Department of Food Science and Technology, University of California at Davis

Donald W. Schaffner, Ph.D., extension specialist, Department of Food Science, Rutgers University

Thanks to the following people and organizations for providing interviews, references, and guidance:

Drs. Battigelli, Bruhn, Cliver, and Shaffner

David K. Bandler, professor of food science, Cornell University

Mindy Brashears, Ph.D., food-safety specialist, University of Nebraska Cooperative Extension

Robert L. Buchanan, Ph.D., chief scientist, President Clinton's Food Safety Initiative

Felicia Busch, M.P.H., R.D., American Dietetic Association

Robert G. Cassens, Ph.D., Department of Animal Sciences, University of Wisconsin at Madison

Caroline Smith De Waal, director of food safety, Center for Science in the Public Interest

Michael Doyle, Ph.D., director, Center for Food Safety and Quality Enhancement, University of Georgia

Walter Faber, Ph.D., professor of biology, Manhattan College
Jesse Greenblatt, M.D., M.P.H., state epidemiologist for New Hampshire
Alice Henneman, M.S., R.D., educator, University of Nebraska Cooperative Extension and editor of *FoodTalk*
Edward S. Josephson, Ph.D., adjunct professor, Department of Food Science and Nutrition, University of Rhode Island
Pat Kendall, Ph.D., professor, Department of Food Safety and Human Nutrition, Colorado State University at Fort Collins
Deborah Klein, president, Agribusiness Solutions, Greensboro, North Carolina
Stephen J. Knabel, Ph.D., food-safety specialist, Pennsylvania State University
Munir Mobassaleh, M.D., Department of Pediatrics, Division of Pediatric Gastroenterology, New England Medical Center
Ken Moore, director, Interstate Shellfish Sanitation Conference
Tom Montville, Ph.D., Daryl Minch, and all the other experts at the Rutgers University Cooperative Extension
Michael T. Osterholm, Ph.D., M.P.H., state epidemiologist, Minnesota Department of Health
Douglas Powell, Ph.D., food scientist, University of Guelph, Ontario, Canada, and administrator of FSNet
Paul W. Sherman, Ph.D., professor, Department of Neurobiology and Behavior, Cornell University

Susan Sumner, Ph.D., food microbiologist, Virginia Polytechnic and State University
Barry Swanson, professor of food science, Washington State University
Donn Ward, Ph.D., professor of food science, North Carolina State University
Experts at the Cooperative Extension Service, University of Georgia

Our deepest gratitude also goes to:

The many people who candidly shared their personal experiences with food-borne infections
Literary agent Judith Riven for all her hand-holding through a tight deadline and for her encouragement when the going got rough
Dell editor Maggie Crawford for her passionate involvement and razor-sharp editing skills
Dear friends Pat R. Gilbert of *The Record*, Bergen County, New Jersey; and Robin Rapport, Jill Cimafonte, Suzanne Barlyn, Joan Plumb, and Judith Norkin-Beyen for their feedback and many words of encouragement
Members of the Mercer County Writers' Collective and Woman's Support Group for their moral support and applause when the manuscript was completed
Zoe Mae and Aaron Joon Levinson for their love and patience; and
Husband extraordinaire, Larry Levinson, for exuding total confidence, for offering insightful criticism, and for taking care of the kids

FOREWORD

Almost fifteen years ago, when I was a young physician training in infectious diseases, I entered a fellowship program at an isolation hospital in Manchester, England. There I saw patients of all ages—up to twenty per week—who were suffering from food-borne diseases of various types—bacterial, protozoa, or viral. Many of these patients were either young or elderly, but every age group was represented. While many patients also had an underlying medical condition that had lowered their resistance to disease, a substantial proportion of adult patients had been perfectly healthy before contracting a food-borne infection.

To arrive at a diagnosis, we tested patients' blood or stool for the presence of the food-borne bacteria or protozoa we knew about. We also relied on our clinical observations such as fever, bloody diarrhea, and abdominal pain. It was often possible to tell if a patient's gastrointestinal infection was bacterial in nature because the odor of their stools was unusually foul.

Most patients who came into our unit had been infected with either *Salmonella* or *Campylobacter*, the two most prevalent food-borne pathogens in the United States today. Some of the children in our unit had cryptosporidiosis, a gastrointestinal disease caused by the water-borne protozoan *Cryptosporidium*. This same bug killed sixty-seven people and sickened more than 400,000 in Milwaukee in 1993. In most cases, food-borne infections resolve themselves in a few days with no medical intervention at all. The patients we saw in Manchester represented exceptions to that rule. Many had become so dehydrated that we needed to administer fluids intravenously. Some patients required antibiotics. Tragically, a few died. But the outcome in the vast majority of cases was good; apart from feeling totally drained, most patients were well on their way to recovery after two or three days in the hospital.

During the eighteen months I spent in Manchester, I saw only one patient infected with *E. coli* O157:H7—the bacterium that later would become infamous as the "hamburger disease" bug. My patient was a young boy with severe bloody diarrhea who required blood transfusions and ultimately developed kidney failure. *E. coli* O157:H7 was rarely encountered in 1986, but any encounter with this bacterium is hard to forget. *E. coli* O157:H7 is especially virulent because it emits a poison known as Shiga toxin. Once the toxin enters the bloodstream, the only thing doctors can do is support the body's vital functions as it tries to heal itself. That was true back in 1986, and it remains true today.

After completing my fellowship, I decided to join the New England Medical Center in Boston because it housed a world-class laboratory populated by experts who were investigating enteric bacteria, infectious agents that attack the digestive tract. At the time, I had not yet decided that Shiga toxins were in my future. I did, however, wish to learn more about enteric bacteria, many of which are food-borne. And I was still haunted by the image of that little boy in Manchester, whom I lost track of after he was transferred to a pediatric renal hospital. As a physician, I was horrified by the pain and suffering experienced by victims of this devastating disease. As a scientist, I was fascinated by the havoc that Shiga toxin played in the human body.

When I first started researching Shiga toxins in 1987, the recently discovered *E. coli* O157:H7 was considered a mere scientific curiosity that occasionally caused disease. This perception changed dramatically when the infamous Jack in the Box hamburger disease outbreak in 1993 catapulted *E. coli* O157:H7 to the front pages of newspapers across America. Of more than 700 people who got sick, fifty developed serious kidney disease, and four young children died. For the first time, the public realized that the consequences of food-borne disease can be far worse than a self-limiting bout of diarrhea and abdominal cramps. Parents in particular were terrified to learn that an active, healthy child could be dead days after eating a contaminated hamburger—a hamburger that looks, smells, and tastes just like an ordinary hamburger.

Soon after this outbreak, many more research groups became interested in *E. coli* O157:H7 and its Shiga toxin, and more private and public research funds became available. Scientists already knew that *E. coli* O157:H7 was but one of many different strains of bacteria that produce a type of Shiga toxin. Indeed, Shiga toxin was the one thing that all these different bacterial strains had in common.

At the New England Medical Center in 1993, my

colleagues and I realized that we already had a head start on Shiga-toxin research; for a number of years we had been working on ways to treat and prevent hamburger disease. In fact, we had already developed tests to detect Shiga toxins and had devised some potential strategies to thwart their deleterious effects in the human body. When the Jack in the Box outbreak occurred, it seemed a logical time to transfer our work in the laboratory to the real world. That led to my collaboration with Meridian Diagnostics Inc., in Cincinnati, Ohio, to develop commercial tests to detect Shiga toxin and toxin-producing bacteria in food as well as in the stools of infected people. The Shiga-toxin test we produced was approved by the Food and Drug Administration in 1995 and is now used in many countries throughout the world.

Although we now had a means of detecting these bacteria and their toxins, the major frustration was, and continues to be, the lack of an effective treatment to prevent the infection's serious complications, such as kidney failure.

In 1997, my colleague, Andrew Plaut, M.D., and I established the Food Safety Initiative at the New England Medical Center. By coordinating discoveries made by clinical and basic science investigators at our institution and elsewhere, we advise researchers, physicians, journalists, industry, and members of the legal profession who need information on the clinical aspects of food-borne diseases. We hope that our efforts will lead to a better understanding of what these vicious microorganisms do to people, for only then will we be able to develop the optimal preventive and therapeutic strategies.

Where do we go from here? The attack on food-borne pathogens is proceeding on various fronts. In addition to learning more about how food-borne microorganisms cause disease, we and other researchers continue our search for therapies to prevent serious complications. We are also developing vaccines against various food-borne pathogens and the toxins they produce, and we are learning more about how bacteria enter our food chain in the first place.

No matter what physicians and scientists come up with, it is consumers who have the greatest power to prevent the spread of food-borne infections. Education is a critical step—and it is the mission of this book. You will learn how these deadly microbes can cause disease and kill. You will learn how to handle and prepare food in ways that minimize the risk of infection. You will learn how to protect yourself when eating out. You will learn how to assess your personal risk of contracting a serious food-borne disease, as not all of us are equally susceptible. While healthy people certainly have become seriously ill from food-borne infections, the very young, the very old, those with compromised immune function, and certain other groups are more likely to die from complications of a food-borne disease.

By the time you finish this book, you will know what, if any, action you need to take if you suspect that you or a loved one has contracted a food-borne infection. Most important, you will know the steps to take, and why you need to take them, to significantly reduce your risk of becoming a victim.

It is my sincere hope that by following the advice laid out in the pages ahead, you will avoid the pain, suffering, and potential tragedy wrought by food-borne disease.

DAVID W. K. ACHESON, M.D.
Director, Food Safety Initiative
New England Medical Center
Boston, 1998

I

"IT MUST'VE BEEN SOMETHING I ATE":

An Overview of Food-borne Illness

Hungry after playing tennis, Cindy L. heads over to a neighborhood restaurant that happens to be a popular hangout for local retirees. On her way in, she passes a group of senior citizens sitting on plastic chairs outside the restaurant door, greeting friends who stream in for a bite to eat. Inside the restaurant, everyone seems oblivious to the fact that their lives may be in danger. The peril isn't an armed robber, a ticking bomb under the cash register, or even a grease fire. The threat is dangerous food-borne microorganisms that are tasteless, odorless, and invisible to the naked eye.

A glance around the restaurant reveals a confluence of human errors that could lead to abdominal pain, diarrhea, vomiting, long-term disability, or even death for anyone who eats at the restaurant that day—particularly the most elderly patrons. Waitresses wipe tables with dirty cloths. Behind the lunch counter, three dozen eggs, which should be refrigerated, are stacked next to the grill. The cook uses the same spatula to flip partially cooked hamburgers and to remove fully cooked hamburgers from the grill. The server nonchalantly sticks her thumb into the pancakes as she delivers them to Cindy's table. To top it all off, the pancakes are slightly raw inside.

Cindy has read about food-borne diseases, and she is outraged by her experience at the restaurant. This book will show you why—from a microscopic point of view—she has every right to be. More important, *Safe Eating* will help you protect yourself and your family from food-borne infections by explaining:

- How bacteria, viruses, and parasites can infiltrate food and drinking water, and which foods are most vulnerable to these biological hazards;
- What government and industry can and cannot do to protect you from food-borne infections;
- Why certain people are at high risk of contracting a food-borne disease;
- Why *Campylobacter* and *Salmonella* have become the most common causes of food-borne illness, and why *Escherichia coli* O157:H7 is among the most deadly;
- Why more virulent strains of bacteria are emerging, intensifying the need for everyone to follow safe food-handling practices;
- How these "bugs" wreak havoc inside the

human body, and which, if any, treatments can help;
- When to seek medical attention, and how to help your doctor diagnose a food-borne disease;
- How to reduce the risk of eating contaminated food at home and in daycare centers, nursing homes, restaurants, and other settings.

As a consumer, you have more control over eating safely than you probably realize. The Centers for Disease Control and Prevention (CDC) estimates that 85 percent of identified cases of food-borne infections stem from two primary sources: failure to keep hot foods hot and cold foods cold, and poor hand-washing practices.

The Scope of Food-borne Illness

Each year in the United States, millions of people experience food-borne infections, but only a tiny fraction of these cases are ever recognized or reported. Since 1980, various researchers have come up with estimates ranging from 1.4 million to 150 million food-borne infections occurring annually across the country. In a May 1996 report, the General Accounting Office concluded that there are 6.5 million to 81 million cases of food-borne illness a year. Bob Howard, special assistant to the director of the National Center for Infectious Diseases, says he has heard figures ranging as high as 300 million cases. Stephen J. Knabel, Ph.D., a food-safety specialist at Pennsylvania State University and the lead author of a scientific status report on food-borne illness prepared for the Chicago-based Institute of Food Technologists, estimates that about one in four Americans suffers a food-borne infection each year. By his assessment, no American family is untouched.

Food-borne illness is so widespread because bacteria and other microorganisms are ubiquitous on Earth. Scientists have gathered bacteria from clouds above mountain peaks and from the deepest depths of the ocean. Bacteria exist in and on animals and people. They are even in the air we breathe. Most of these microorganisms are harmless, or even helpful; some are not.

"Part of the problem is that we don't really understand food-borne diseases in this country. We only understand bits and pieces," says Dr. Michael T. Osterholm, Ph.D., M.P.H., Minnesota state epidemiologist and a nationally known authority on food-borne infections. A survey conducted by Osterholm's department in 1997 indicated that there were probably 7.8 million episodes of diarrheal illness in Minnesota that year. However, fewer than 2,000 cases of *Salmonella, Campylobacter, E. coli, Shigella,* and *Yersinia* infections were reported to the state health department. While diarrheal diseases can be triggered by numerous factors, Osterholm says a substantial proportion of these episodes are probably caused by food-borne microorganisms. Not counted by the Minnesota survey were food-borne infections that create symptoms other than diarrhea, and illnesses caused by food-borne microorganisms that no one is looking for.

Osterholm cites other studies suggesting that stomach illnesses are on the rise in the United States. One conducted in Cleveland, Ohio, between 1948 and 1957, and a study done in Tecumseh, Washington, between 1965 and 1971, both found an average of one stomach illness per person per year. But studies in five cities around the United States done in the late 1990s found a rate of 1.4 stomach illnesses per person per year. In Minnesota, where food-borne illness has been more intensely investigated, the rate was even higher—1.8 illnesses per person per year. It is not uncommon for multiple cases of food-borne illness to occur in a single individual who is in a high-risk group and who does not take proper precautions when handling food. A similar pattern can be seen among children who share food or eating utensils in daycare, school, or summer-camp settings.

The Centers for Disease Control and Prevention (CDC) receives reports of about 500 food-borne disease outbreaks annually involving a total of about 20,000 people, which represents just the tip of the iceberg.

The vast majority of food-borne illnesses resolve themselves without treatment after a few days, but a significant number of victims develop such serious complications as kidney failure, arthritis, and paralysis. *Campylobacter jejuni,* which frequently contaminates raw or undercooked poultry, has become the most common cause of nonaccident-related paralysis in the United States. *E. coli* O157:H7 infection is now the leading cause of acute kidney failure in children. According to one mathematical model endorsed by the National Center for Health Statistics, up to 9,100 Americans die of a food-borne infection each year. That figure translates to about twenty-five fatalities a day, on average.

"I think most people believe that food-borne illness is just temporary discomforts, and they don't realize the seriousness of other possible consequences," says Christine M. Bruhn, Ph.D., director of the Center for Consumer Research at the University of California at Davis.

The good news is that food-borne illnesses are often preventable. "Fight BAC! The Partnership for Food Safety Education," a consortium of industry, government, and scientific groups, recommends that consumers follow four key steps—clean, separate, cook, and chill. These and many other food-safety precautions are elaborated upon throughout this book.

The information we are presenting about food safety and food-borne illnesses is based on the latest scientific and medical research and on the insights and recommendations of dozens of food-safety experts working in academia, government, and industry. These experts include biologists and microbiologists, public health specialists, home economists, food-safety educators, epidemiologists, consumer advocates, physicians, food technologists, food industry consultants, researchers, nutritionists, infectious disease specialists, and officials from the Centers for Disease Control and Prevention, and officials from the Food and Drug Administration (FDA) and the United States Department of Agriculture (USDA), the two agencies that regulate the production and processing of almost all of the food sold in this country. In addition, victims of food-borne infections candidly share their experiences in the hope of helping others avoid similar misfortunes.

Safe Eating is divided into three parts. In Part 1 we

explain how food-borne microbes can make people sick. In addition to presenting the possible short- and long-term medical consequences of food-borne infections, we explain how physicians diagnose and treat these diseases. We also show how individual cases fit into the overall public-health scheme. Except for the scenarios we created to illustrate medical facts in Chapter 2, the anecdotes included in this book represent actual cases, although names have been changed or surnames omitted to protect people's privacy.

Part 2 takes an in-depth look at how meat and poultry, seafood, produce, eggs and dairy products, and drinking water can become contaminated by disease-causing microbes; how the government and industry are tackling the problem; and how consumers can make the wisest food choices possible. Some of the information about contamination routes is very explicit and may be disturbing to some readers. We have included it here because we believe it is an essential part of your food-safety education and that it will motivate you to take extra care when you buy, store, prepare, and serve food. Each chapter in this section includes a summary of what consumers can do to protect themselves.

Part 3 offers a cornucopia of general and specific food-safety tips and guidelines that have been culled from a wide variety of authoritative sources.

The final chapter, "Challenges Ahead," presents information about the future of food safety and the importance of food-safety education. Resources to help you stay abreast of the fast-paced world of food safety appear in Appendix A.

Growing Threats

The importance of following proper food-handling practices has never been greater. Two decades ago scientists were aware of about seven microorganisms that could cause food-borne illness. Today there are at least eighteen bacteria, three viruses or virus groups, ten protozoa (single-celled animals), and five toxins known to cause food-borne disease in humans. Practically every week, another food-borne illness outbreak, suspected outbreak, or food recall makes headlines. Food products historically thought safe, such as eggs, raw fruit juice, even breakfast cereal, have become potential vehicles for infectious disease. In June 1998, Malt-O-Meal Incorporated recalled 2 million to 3 million pounds of plain, toasted oat cereals because of a possible link to an outbreak of *Salmonella agona* infection involving at least 188 people in eleven states. Forty victims were hospitalized. The source of the contamination, possibly a faulty oven that allowed a vitamin spray to spoil the cereal, was under investigation at this writing. Regardless of that investigation's outcome, the outbreak "points to the ability of these critters to survive on a dry surface," says Robert E. Anderson, Ph.D., professor emeritus in the Division of Plant Sciences at West Virginia University. "Drying is analogous to freezing in that the product gets dehydrated. *Salmonella* has been found to live for years in river ice. It just shuts down all its metabolic processes, produces antistress proteins, and copes."

Another problem is food-borne bacteria that are evolving into more virulent strains. One example is *E. coli* O111, which was recently detected in Italy and

Australia in patients suffering from hemolytic uremic syndrome (HUS), a potentially fatal kidney disease that occurs as a complication of certain kinds of food-borne illness. In the United States, HUS is more closely associated with *E. coli* O157:H7 infection (also known as hamburger disease). While these two strains, or serotypes, have different physical characteristics, both have the ability to produce the same poison called Shiga toxin, which can lead to HUS. Scientists believe that *E. coli* cells are routinely exchanging genetic information, in this case the Shiga-toxin gene. This genetic exchange can produce a relatively benign bug that causes mild diarrhea, or a lethal bug that causes HUS and death. Scientists don't know if *E. coli*'s gene chopping and swapping is happening more frequently now than it did in the past, or if it just seems that way because better tools are now available to detect these hybrid bacteria.

Another relatively recent change has been in the characteristics of food-borne illness outbreaks. In the past, most outbreaks were fairly obvious, usually traced to picnics, weddings, church suppers, and other gatherings where highly contaminated food was served, and the victims tended to live in the same geographic region. Over the past two decades, outbreaks of food-borne illness have become more frequent, more diffuse, and in many cases, more difficult to identify for a variety of reasons:

- Food production is more centralized than it used to be. Eggs from one processing plant in Ohio, or beef from a single slaughterhouse in Texas may be consumed in twenty or more states. If these eggs or beef products are contaminated at their source and eaten undercooked, the results would be seemingly unconnected, or sporadic, cases of illness. It is unlikely that the same doctor would see more than one infected patient.

- A contaminated egg, piece of beef, or chicken part can contaminate an entire batch when these products are pooled or ground together, spreading low-level food-borne contamination far and wide. An innocuous level of bacteria can amplify to a dangerous dose if the food's temperature is not properly controlled. Some strains of food-borne bacteria and protozoa have a very low infectious dose; just a few cells may be all it takes to make someone sick.
- The number of families in which both parents work outside the home has risen sharply, leaving mothers and fathers with less time to devote to shopping and meal preparation, and more children assuming responsibility for these tasks.
- Americans travel and eat out more often than they did in the past, spending almost half of all their food dollars in food-service establishments. Each time you patronize a restaurant, you place your trust in food-service workers who may be inadequately trained to handle food safely.
- Increasing numbers of Americans are demanding foods that are organically grown, unprocessed, or minimally processed. While these kinds of foods offer many health benefits, they can be easily contaminated with microorganisms if precautions are not taken. (Proper washing and handling of

fresh produce in most cases eliminates or reduces any pathogens to harmless levels.)

- Americans have demanded, and come to expect, all types of raw produce to be available twelve months a year. This demand is met by importing fruits and vegetables from countries where food safety may not be as well controlled as it is in the United States.

Safe Processing

Many people assume that raw, unprocessed foods are safer and more healthful than processed foods, but this is not always the case. Almost every form of food processing, including pasteurization, canning, curing, and irradiation not only extends shelf life by thwarting the growth of spoilage bacteria, but also makes foods safer. As Tom Montville, Ph.D., professor of microbiology and chairman of the Department of Food Science at Rutgers University, points out, "canned," "frozen," and "processed" are not opposites of "fresh." The opposite of "fresh" is "spoiled."

For instance, the food additive nitrite prevents botulism in bacon and other cured, vacuum-packed meats. Botulism is caused by a potentially lethal toxin produced by the bacterium *Clostridium botulinum,* which thrives in an air-free environment.

Another example is frozen vegetables, which you may presume contain fewer vitamins and minerals than their nonfrozen counterparts. Research shows, however, that the opposite is usually true. "In almost every case, frozen vegetables are more nutritious," says Felicia Busch, M.P.H., R.D., a spokeswoman for the American Dietetic Association. Busch says that frozen-food processors use only the strongest, heartiest vegetables, which by definition are nutrient-rich. Fresh vegetables may arrive at your local grocery store days or weeks after they were harvested. As time passes, raw vegetables lose nutritional value. Bumping and bruising of raw vegetables during packing and distribution lead to further nutrient loss. If vegetables are frozen in the field during harvest, as many packages of frozen vegetables are, nutrients get locked in. Frozen vegetables also have less opportunity to pick up contamination from food handlers en route to the consumer. Any bacteria on the vegetables before they were frozen will not grow in a freezer, but bacteria can grow and multiply at higher temperatures.

Fresh, raw juice has also traditionally been considered a health food. "But I continue to think of the family whose sixteen-month-old daughter was given a drink of nice, fresh fruit juice that had been refrigerated, and that child ended up in the hospital and in the grave," says Bruhn of the Center for Consumer Research. She is referring to Anna Grace Gimmestad of Colorado, who died in 1997 after drinking unpasteurized Odwalla apple and fruit juice blend contaminated with *E. coli* O157:H7. At the time, Odwalla said it hadn't realized that this dangerous bacterium could survive in something as acidic as apple juice. The public didn't realize it either. "But," Bruhn points out, "that information had been in the professional literature for more than ten years." The company and the parents apparently were so excited about the benefits of fresh juice, she says, that they forgot about the benefits of pasteurization. Odwalla now pasteurizes its juice products, and food-safety experts continue to strongly advise consumers—especially children, the elderly, and those with weakened immune systems—to avoid unpasteurized juice.

Growing Public Concern

Highly publicized tragedies like Anna Gimmestad's help fuel public concern about microbiological hazards in foods. A periodic survey sponsored by the Food Marketing Institute asks respondents to name their biggest food-safety concern. In 1993 bacterial contamination was mentioned by 46 percent of respondents; by 1997 that figure had jumped to 69 percent. Bruhn says the professional food scientist would concur with these concerns. "People becoming ill and people dying from microbiological hazards is a real safety concern, and it's appropriate that the public pay attention to these issues."

In recent years the federal government has paid more attention to food safety as well. The latest series of federal food-safety programs was touched off by the December 1992 outbreak of *E. coli* O157:H7 infections among people in the Northwest who ate undercooked hamburgers at Jack in the Box, a fast-food restaurant chain. More than 700 people became ill and four children died. In 1993 the USDA began requiring raw meat and poultry products to carry safe-handling labels that address storage, cooking, and holding practices. The next year, the USDA declared *E. coli* O157:H7 in ground beef an "adulterant" (an illegal contaminant) and launched a monitoring program for this bacterium in ground beef. In 1995 the USDA, FDA, and the Centers for Disease Control and Prevention (CDC) initiated a food-borne illness surveillance program, now called FoodNet, to collect better information on the incidence of food-borne infections, including those caused by *Salmonella* and *E. coli* O157:H7.

The centerpiece of the government's crackdown on food-borne illnesses is a food-safety system with a complicated moniker: "Hazard Analysis and Critical Control Points," commonly known as HACCP (pronounced "HASS-ip"). HACCP requires food processors to identify the steps along the food-processing continuum where contamination is most likely to occur and then put in place controls geared toward reducing or preventing contamination.

Specifically, HACCP involves seven steps:

1) **Identify hazards.** These could be biological (i.e., bacteria or virus), chemical (i.e., mercury), or physical (i.e., ground glass or metal).
2) **Identify critical control points.** These are phases in a food's production—from its raw state through processing to consumption—at which the potential hazard can be controlled or eliminated (i.e., cooking, chilling, handling, cleaning, and storage).
3) **Establish preventive measures with "critical limits" for each control point.** For a cooked

product, critical limits would include temperature and time (i.e., the temperature at which the product must be heated and for how long).

4) **Establish procedures to monitor the control points.** This might include determining which employees are responsible for monitoring cooking time and temperature.
5) **Establish corrective actions to be taken when monitoring shows that a critical limit has not been met** (i.e., reprocessing or disposing of food that was not cooked at a temperature high enough to kill bacteria).
6) **Establish effective record-keeping methods to document the HACCP system.**
7) **Establish procedures to verify that the system is working consistently** (i.e., testing time- and temperature-recording equipment on a periodic basis).

HACCP was created three decades ago for the space program to ensure that the astronauts' food would be safe. Starting in the late 1970s, the FDA issued HACCP-based regulations for canners. In the mid-1990's, both the FDA and USDA began phasing in HACCP rules for other segments of the food industry, including seafood, meat, and poultry operations where the old "poke-and-sniff" visual inspection method had reigned supreme for ninety years. A cottage industry of HACCP consultants and microbial testing firms has grown up around this new mandate.

In April 1998 the Federal Food and Drug Administration announced it was recruiting restaurants, grocery stores, institutional food service establishments, and vending operations to voluntarily integrate HACCP into recipes and standard operating procedures. According to the FDA, these retail HACCP pilot programs "will look at each establishment's role in continuous problem solving and prevention, rather than relying on periodic facility inspections by regulatory agencies."

In a simplified way, consumers can use HACCP principles in their own kitchens. For example, in Chapter 9 you will find lists of cooking temperatures and storage limits for various foods, as well as instructions on calibrating and using a meat thermometer. Grocery shopping techniques to optimize food safety also are included.

Food Safety at Uncle Sam's Table

At this writing, President Clinton has proposed a 13 percent increase for federal food safety and inspection programs in the 1999 fiscal-year budget, bringing the total budget to about $900 million. Among other things, the money will be spent on research, testing of meat and poultry for *Salmonella* and *E. coli* O157:H7, hiring and training of more food-safety inspectors, food-safety education programs for the elderly, monitoring farming and food-production operations in countries that export food to the United States, and improving surveillance of food-borne illnesses.

Despite the many positive steps taken by the federal government, there have been some missed opportunities. For example, many food-safety experts want the government to push industry harder to use irradiation, also known as radiation pasteurization. This so-called "kill step" exposes foods to low doses of gamma rays, X rays, or beams of electrons, all of which drastically reduce or eliminate pathogens and extend shelf life by killing spoilage bacteria. Many people recoil at the concept of food irradiation because they associate radiation with sickness, birth defects, and environmental contamination. But exhaustive scientific research shows that the process can be done safely and that it does not leave food radioactive. Studies also indicate that once consumers are educated about the process, the majority are interested in buying irradiated foods. Nevertheless, very few foods are currently being irradiated in this country. The reasons for this are discussed at various points throughout this book.

Another missed opportunity occurred during the August 1997 nationwide recall of 25 million pounds of Hudson Foods ground-beef patties suspected of being contaminated with *E. coli*. Food-safety specialist Stephen Knabel says that that was an "excellent time for everybody to step up and educate consumers" about safe handling and cooking procedures for ground beef. "But the consumers were not educated at all," he contends. "In fact, consumers were misinformed about food safety. The government left the consumer with the impression that it's up to the producers and processors to totally eliminate pathogens from raw animal foods."

In reality, everyone along the food supply continuum—from the farmer and fisherman to the consumer—plays a pivotal role in reducing the risk of food-borne illness.

Myths and Facts About Food Safety

As you progress through this book, you may be surprised to learn that some of your long-held assumptions about food safety are wrong. To get you started, here are some common food-safety myths and facts. The first three come from the February 1998 issue of *FoodTalk,* an award-winning electronic newsletter written by Alice Henneman, M.S., R.D., of the University of Nebraska Cooperative Extension in Lancaster County (subscription information appears in Appendix A). The rest were adapted from a column in the April 1998 issue of *Food Technology,* published by the Institute of Food Technologists. Author Christine Bruhn of the Center for Consumer Research says she wrote the column to correct myths she found in recently published cookbooks.

MYTH #1: "If it tastes okay, it is safe to eat."
FACT: If you trust your taste buds to detect unsafe food, you may be in trouble. Unlike spoilage bacteria that make food look and feel rotten and produce a foul odor, disease-causing microorganisms do not alter the taste, texture, or smell of food. While you can usually rely on taste, smell, or sight to determine if something is fresh, you must rely on your knowledge and intellect, not your senses, to figure out if something is safe to eat.

If food has begun to rot, there is a chance that pathogenic bacteria have also multiplied. So taking even a tiny bite to test the freshness or safety of a questionable food can be dangerous. For some food-borne illnesses, such as botulism, eating just a small amount of contaminated food can be fatal.

MYTH #2: "We've always handled our food this way, and nothing has ever happened."
FACT: If you use past experiences to predict whether a food is safe, your future may include a food-borne illness. Many incidents of food-borne illness went undetected in the past. Food-borne illness symptoms of nausea, vomiting, cramps, and diarrhea were often and still are erroneously blamed on the "flu." Also, both the nature of our food supply and the virulence of food-borne pathogens have changed. For example, in the past, the chicken served at dinner might have been walking around during breakfast. Today, your food may travel halfway around the country or the world before it arrives at your table. Because most food passes from producer to processor to trucker to retailer before it reaches you, the opportunities for mishandling are greater than they were in the past.

More potent forms of bacteria present further problems. For example, in 1990 the U.S. Public Health Service cited *E. coli* O157:H7, *Salmonella, Listeria monocytogenes,* and *Campylobacter jejuni* as the four most serious food-borne pathogens in the United States. Twenty years ago, three of these—*Campylobacter, Listeria,* and *E. coli* O157:H7—weren't even recognized as sources of food-borne disease.

MYTH #3: "I sampled it a couple of hours ago and didn't get sick—so it should be safe to eat now."
FACT: Though you may feel all right a few hours after eating a food, the food still may be unsafe for you and others to consume. A food-borne illness may develop within an hour or a few days of eating contaminated food; symptoms of food-borne illness can also occur as long as two or more weeks later. If sickness occurs more than twenty-four hours after eating a food—which is often the case—it is frequently blamed on other causes.

Another consideration: While one person may eat a food and not get sick from it, someone else may be more susceptible to food-borne illness because his or her immune system is weaker or less mature. Youth, old age, certain acute or chronic illnesses, and certain medications all make people more vulnerable when exposed to food-borne pathogens.

MYTH #4: "Organic produce does not need to be washed."
FACT: All produce is susceptible to dirt, insects, and harmful microorganisms; thus, all fruits and vegetables, regardless of how they were grown, should be carefully washed before they are eaten.

MYTH #5: "Produce should be washed with water and a few drops of pure soap."
FACT: Washing food with soap or detergents is not recommended by the FDA or USDA because

little is known about the health effects of eating soap or detergent residues. Moreover, it is not known if soap is effective in ridding pesticide residues from produce.

MYTH #6: "Modern animal production has reduced the quality and safety of meat and poultry."
FACT: Raw animal foods have always had the potential to contain pathogenic bacteria, regardless of where or when the animals were raised.

MYTH #7: "It is okay to use raw eggs in no-cook recipes."
FACT: Use of raw eggs poses a safety risk, which is greater for people whose immunity to infectious diseases is compromised.

MYTH #8: "A cooking temperature of 155°F is high enough to kill *E. coli* O157:H7 in ground beef."
FACT: A cooking temperature of 155°F may kill *E. coli* O157:H7, depending on the cooking time at this temperature and on the bacteria levels in the beef. However, for a greater margin of safety, the USDA and FDA recommend a cooking temperature of 160°F.

MYTH #9: "Buying top-grade beef, grinding it to order, or home grinding it can reduce the risk of *E. coli* O157:H7 contamination."

FACT: No studies support these recommendations, and home grinding increases the opportunity for "cross-contamination." Cross-contamination occurs when pathogens are transferred by hands, cutting boards, or other food-contact surfaces to food that is ready to eat.

Of course no amount of information can guarantee that you will never suffer a food-borne illness; many of the food-safety experts interviewed for this book have been infected. You can, however, greatly reduce your risk for future infections by following the guidelines laid out in *Safe Eating.* Food should not be a threat; it should remain one of life's greatest pleasures. Armed with food-safety knowledge, you will have the power to keep it that way.

PART I

MEDICAL ASPECTS OF FOOD-BORNE DISEASES

2

SICK TO YOUR STOMACH:

How "Bad Bugs" in Food Make Us Ill

The following scenario may seem familiar: Tom is in a hurry. It is two o'clock on a Saturday afternoon, and his daughter's ballet performance begins in thirty minutes. The theater is a twenty-minute drive away. Tom has not had lunch, and he's famished. So he grabs a semifrozen chicken breast from the refrigerator and pops it into the microwave for three minutes. He puts the chicken on a roll, jumps into his car, and eats the sandwich en route to the performance. After he takes his first bite, Tom notices a little pink spot on the chicken breast but wolfs it down anyway because the chicken smells and tastes fine.

Unbeknownst to Tom, inside that pink spot, approximately 10,000 *Salmonella typhimurium* cells are stirring out of cold-induced dormancy. *S. typhimurium* is one of 2,213 known strains of *Salmonella* bacteria, many or all of which are potentially harmful to human health, and more serotypes are being identified all the time. For *Salmonella enteritidis,* the bug implicated in a 1994 outbreak of gastrointestinal illness caused by contaminated ice cream, the infectious dose appears to have been no more than twenty-eight cells, according to researchers.

Each sausage-shaped bacterium sports a tiny tail called a flagellum (pronounced "fla-GELL-um) and a halo of hairlike structures called "pili" that whip and thrash about, enabling the organism to move. The bacteria were born in the digestive tract of the chicken Tom ate when it was alive on a farm eight hundred miles from his home. The bacteria were excreted in the chicken's feces, some of which clung to the bird's tail feathers as it was trucked to the slaughter plant. During slaughter and processing, as frequently happens, a microscopic amount of fecal material found its way into the bird's breast tissue. Various efforts were made in the plant to kill or remove all fecal material from the carcass, but some of the bacteria stuck stubbornly to the muscle. There it remained as the chicken was deboned, packaged, and transported to Tom's local supermarket.

Four days after it arrived at the supermarket, Tom's wife purchased the package of chicken breasts and put it, along with her other groceries, in the trunk of her car. On her way home that warm spring day, she stopped at the cleaners and then at the photocopy center. The temperature inside the trunk climbed to 100°F, heating the chicken breasts to 58°F—warm enough for *Salmonella* to grow and divide. When she got home, she put the contaminated chicken into the freezer, where the bacteria remained alive but dormant. Two weeks later—the day before the ballet performance—Tom's wife moved the chicken to the refrigerator, expecting to cook it over the weekend.

Tom's undercooked chicken sandwich served as a temporary shelter for *Salmonella,* but his intestine provided the microbial homestead for them. The human gut is chock-full of things bacteria require for growth and proliferation: nitrogen, oxygen, carbon in the form of carbohydrates, and amino acids, which are needed to manufacture proteins. Nevertheless, only the heartiest of the bacteria Tom ingests make it past his stomach. Had his *Salmonella* "dose" been somewhat lower, the acid secreted by his stomach would have wiped out the whole lot of them. Unfortunately for Tom, several thousand *S. typhimurium* cells survive. Surrounded by food particles, the bacteria slowly squiggle down to his small intestine. Eventually they settle in near the juncture of his small and large intestines—a safe haven in which the bacteria are no longer threatened by the sting of peptic acid.

The intestines can be thought of as a series of barriers designed by nature to allow food molecules in and to keep disease-producing microbes out. For instance, the lower parts of the intestines are protected by such factors as antibodies and mucus. But the principal defense throughout the entire intestinal tract is the closely knit layer of cells lining the intestinal wall. These epithelial cells form a tight barrier that physically prevents most bacteria from invading tissue and blood. But *Salmonella* and other food-borne pathogens have figured out a way to overcome that barrier. They do it with a series of mechanisms that initiate "adherence," in which each bacterium physically attaches itself to the surface of an epithelial cell. Different food-borne pathogen species have different methods of adhering to the host cells; precisely how this occurs is not completely understood. Researchers suspect that certain proteins on the surface of the bacteria have an affinity for molecules on the epithelial cell surface.

The next step is "invasion," in which the bacterium gains entry into the cell that is forming the barrier. An interaction between these bacterial and host components somehow tricks the host-cell wall into letting the bacteria through. The bacteria then multiply and spread to neighboring cells.

Thirteen hours after Tom ate the chicken sandwich, the army of microscopic invaders has grown to more than two million strong. Cells of the epithelial lining begin dying in droves as a result of being penetrated by the *Salmonella* bacteria. Almost immediately, Tom's immune system senses this insult and begins mobilizing its defense forces. Antibodies specially designed to kill only *S. typhimurium* are being churned out by the millions and are closing in on the bacterial colony. This phase of the immune response triggers local inflammation of the intestinal lining, which in turn triggers many of Tom's initial symptoms.

His first overt sign of trouble comes twenty hours after eating his contaminated sandwich when he develops a headache. An hour or so later, he experiences some mild gastrointestinal distress. This symptom gradually worsens until Tom is doubled over with severe abdominal pain caused by intestinal spasms. The large and small intestines do not contain regular nerve endings, like the ones in the skin, so they do not feel pain in the conventional sense. This is why a surgeon can often perform a tissue biopsy in the colon (part of the large intestine) without using an anesthetic. The intestines, however, are extremely sensitive to trauma by toxins or bacteria that don't belong there. And the intestines respond to these insults by squeezing like crazy. All this squeezing makes Tom's abdomen feel extremely uncomfortable. And his cramps are worsened by the inflammation brought on by the immune system cells fighting the invaders.

Suddenly Tom is gripped by an intense urge to use the toilet. He sprints to the bathroom and endures a painful, ten-minute bout of watery, explosive diarrhea. Overcome by a wave of nausea, Tom throws up, too. Not everyone who gets *Salmonella* infection develops a fever, but Tom does; the thermometer reads 100.5. Assuming he has the flu, he crawls into bed, only to wake up every thirty to sixty minutes for another episode of vomiting and violent diarrhea. This pattern continues throughout the next day. His appetite gone, Tom can eat nothing for the rest of the weekend. By Monday Tom's bathroom visits have become much less frequent, but he calls in sick because he is too weak and exhausted to work.

Tom has salmonellosis, one of the most widespread food-borne infections in the United States. Most commonly spread through raw or undercooked eggs, *Salmonella* can also exist on most animal food products, especially chicken.

Tom became sick even though, as a healthy, robust man, he doesn't fit into one of the "high-risk" groups for food-borne disease. If his immune system had been weak, the *Salmonella* may have had enough time to invade deeper, into the actual wall of his intestine as well as into his bloodstream. This would have caused bacteremia, commonly known as blood poisoning. Bacteremia can be a serious illness marked by high fever, chills, headache, rapid breathing, and disorientation. If the amount of bacteria and their toxins in the bloodstream is high enough to cause tissue damage and a dramatic drop in blood pressure, the condition is called septic shock. This can be fatal. If Tom had had a scarred or abnormal heart valve, bacteremia could have led to endocarditis, an inflammation of the internal lining of the heart.

Another possible long-term complication of salmonellosis, as well as *Campylobacter, Shigella,* and *Yersinia* infections, is reactive arthritis, a painful joint inflammation that usually affects the knee or ankle. Reactive arthritis may set in three weeks to two months after the acute stage of the disease. This problem may actually be fallout from the body's immune response to the food-borne infection. During the disease's initial stage, the immune system manufactures antibodies that destroy the pathogens. Once the pathogens are gone, these antibodies sometimes attack the body's own tissue in the joints, in what is known as an autoimmune response. This results in inflammation and pain, which may be chronic or transient.

In people with a certain genetic predisposition, salmonellosis can lead to Reiter's syndrome, which is characterized by conjunctivitis (an eye infection), and arthritis later on. It is estimated that food-borne *Salmo-*

nella causes 2 to 4 million illnesses every year in the United States. The fatality rate for most forms of salmonellosis is less than 1 percent.

If the pathogen Tom ingested happened to be *E. coli* O157:H7, he might have developed more serious complications, most notably hemolytic uremic syndrome (HUS) and kidney failure.

Before we continue our explanation of how the most common food-borne bacteria in the United States can make people sick, it is worth noting that food-borne bacteria can be placed in three broad categories according to the mechanism by which they cause human disease:

1) Bacteria that damage the intestine directly (i.e., *Salmonella, Campylobacter, Shigella, Yersinia,* and *Listeria*).
2) Bacteria that make toxins after they have entered and colonized the gut. The most infamous of these is *E. coli* O157:H7, which produces Shiga toxin.
3) Bacteria that make a "preformed" toxin in food (i.e., *Clostridium perfringens, Staphylococcus aureus, Bacillus cereus,* and *Clostridium botulinum*). Illness caused by ingesting preformed toxin (or naturally occurring toxin, as in poison mushrooms) are the only true forms of "food poisoning." However, the term food poisoning is commonly used to describe any food-borne infection.

Some food-borne bacteria cause symptoms by more than one mechanism. And while people can develop an immunity to a particular strain of bacteria after an initial exposure, they may not develop an immunity to a preformed toxin.

Escherichia coli O157:H7

Due to the infamous Jack in the Box outbreak of December 1992 and the hamburger-patty recall by Hudson Foods in 1997, most Americans are now well aware that *E. coli* O157:H7 can lurk in raw or undercooked ground beef. However, as the next scenario illustrates, even vegetarians can be exposed to this noxious organism and can develop what is commonly known as hamburger disease.

Three-year-old Lily loves apples, and her mother, Fran, is delighted, because she and her family are strict vegetarians. Every fall, Fran takes Lily to an orchard at a nearby farm, where they pick apples by the bagful. Fran likes this particular farm because all the produce is organically grown, with no chemical pesticides. She wants the food her children eat to be pesticide-free and as fresh as possible. As an extra measure of protection, Fran makes it a policy to scrub the apples with a stiff vegetable brush under running water before eating them. Fran has read magazine articles about food safety. She feels very much in control.

During one of their apple-picking excursions, Fran lets Lily wander a few yards away to gather her own little stash. In the distance, a deer is grazing among the apple trees. Straining to reach a cluster of apples on a branch, Lily doesn't notice the deer; neither does her mother. The girl looks down and sees an apple on the ground. She picks up the fallen fruit, takes a few bites, tosses aside the remainder, and finally runs to her mother, whose back had been momentarily turned. Lily does not mention having eaten the apple. Even if she had told someone, it would have been too late to prevent infection. Seventy-five *E. coli* O157:H7 cells on the apple skin are already making their way down Lily's esophagus to her stomach.

The infectious dose of *E. coli* O157:H7 is very small, probably between fifty and one hundred cells, and perhaps as low as ten. Therefore unlike some other pathogens, amplification of *E. coli* O157:H7 in food is not essential to cause illness.

In addition to raw or undercooked ground beef, other foods implicated in *E. coli* O157:H7 outbreaks have included roast beef, salami, raw milk, improperly processed apple cider, contaminated water, vegetables grown in untreated cow manure, and cantaloupe. Lettuce and other salad items have also been contaminated with this bug through mishandling. But how could *E. coli* O157:H7 get on an organically grown piece of whole fruit that no one but Lily has ever touched? If you analyze the soil where the apple had fallen, it would reveal a recent visit by that wayward deer. While grazing under the tree that produced Lily's apple, the animal had defecated. Its fecal material contained many bacterial species, among them *E. coli* O157:H7, which is harmless to the deer but potentially life-threatening to a human, particularly a tyke like Lily.

What makes *E. coli* O157:H7 so dangerous is not the bacteria themselves but the Shiga toxin that they produce. The gene that codes for Shiga toxin probably originated in another food-borne pathogen, *Shigella.* It is theorized that the Shiga-toxin-producing *E. coli* O157:H7 was spawned in the 1970s during a dysentery epidemic in Central America. According to this theory, a virus in the process of reproducing acquired the toxin-making gene from *Shigella* and transferred the gene to a harmless strain of *E. coli*—creating the monster we know today. Shiga toxins have since been associated with fifty to sixty other *E. coli* strains and other food-borne bacteria, such as *Citrobacter* and *Enterobacter,* but to a much lesser degree.

Shiga toxin, like many organic poisons, is a protein. If you purify and dehydrate Shiga toxin in a laboratory, you end up with a white, fluffy powder. Reconstitute it with a little water, and the toxin becomes a clear, odorless liquid. If you were to drink that liquid, you probably wouldn't suffer much damage; most of the toxin would be denatured, or degraded, by your stomach acid. If, however, you were to inject Shiga toxin directly into your bloodstream, you'd be in big trouble. Just like Lily.

Forty-eight hours after eating the apple she picked up from the ground, Lily starts feeling queasy. That's how long it took for the *E. coli* O157:H7 to colonize and adhere to her intestinal epithelial cells, multiply, and produce high levels of Shiga toxin. Lily's next symptoms are severe abdominal cramps, watery diarrhea, and nausea. Fran assumes her daughter has a stomach virus and keeps her home from nursery school—even though, curiously, the child has no fever. Fran has nursed two older children through bouts of diarrhea and expects that Lily's episode will also pass without fanfare. Within a few hours, though, Lily's stools turn bloody. Fran has never seen this before in a child. She panics and rushes Lily to the pediatrician. Bloody diarrhea, cramps, and occasional vomiting continue for two more days. The doctor admits Lily to the hospital.

Lily's stools are bloody because the *E. coli* O157:H7 cells are producing enough Shiga toxin to damage the small blood vessels in her intestines and to destroy massive amounts of tissue. In roughly 90 to 95 percent of *E. coli* O157:H7 cases, the body's natural defense system destroys the pathogens in the intestine before serious complications emerge. The natural flora of the intestinal tract also help fight pathogenic bacteria by hogging most of the available nutrients. But Lily has been taking a wide-spectrum antibiotic for an ear infection for almost a week. The drug has wiped out most of the beneficial bacteria in her intestines, creating less competition for the bad guys. Eventually, large amounts of Shiga toxin enter Lily's bloodstream, possibly through the tiny blood vessels that feed her intestinal wall. Doctors suspect that Shiga toxin may also occasionally reach the blood through the urinary tract.

Scientists are trying to figure out exactly how Shiga toxin gets from the *E. coli* to the intestinal tissue and into the blood. In test tubes, the toxin is released only when the bacteria die. Inside the intestine, close proximity to human cells may literally force the Shiga toxin out of a living bacterium.

Shiga-toxin molecules circulate through all of Lily's internal organs and tissues, on the lookout for cells they are capable of binding to. For the most part, those cells are located in the kidneys. Five days after Lily ate that tainted apple, Shiga-toxin molecules have damaged enough kidney cells to cause measurable changes in her urine output. Blood clots begin to form in the kidneys, possibly because the toxin is also damaging the blood vessels feeding these organs. Without adequate kidney function to cleanse her blood, waste products begin to build up in Lily's circulatory system. Dialysis is needed to save her life. Her doctors fear that more blood clots will form elsewhere in her body, including the brain, which could lead to brain damage, strokes, seizures, and possibly death.

Lily is suffering from hemolytic uremic syndrome (HUS)—the leading cause of acute, or sudden kidney failure in children. Of the estimated 20,000 Americans who get sick from *E. coli* O157:H7 each year, approximately 500 die from HUS complications, most of them children. HUS has also been linked to *Enterobacter, Shigella,* and other bacteria, as well as to the AIDS virus, but these associations have been extremely rare. Although *Enterobacter, Shigella,* and other bacteria also make Shiga toxins, there is probably some other mechanism causing HUS in some of these non–*E. coli* cases.

Lily eventually recovers from her illness and is taken off dialysis. But as happens in one third to one half of HUS survivors, there is partial but permanent damage to Lily's kidneys. She may lead a perfectly normal life. Or this chance childhood encounter with *E. coli* O157:H7 could come back to haunt her twenty or thirty years from now in the form of high blood pressure or kidney failure. It is impossible to know for sure.

Campylobacter jejuni

It is equally difficult to predict the long-term consequences of campylobacteriosis, the disease caused by *Campylobacter jejuni,* another food-borne bacterium. According to the USDA's Food Safety and Inspection Service, *Campylobacter* is now considered the most common cause of food-borne illness in this country, with *Salmonella* running a close second.

Campylobacter can exist in the intestinal tracts of people and animals without causing illness if the host has built up an immunity to it. If, however, a person without immunity to *Campylobacter* consumes as few as 500 live *Campylobacter* cells through unpasteurized milk, untreated water, or undercooked meat or poultry, he or she may become extremely ill. (By comparison, the anthrax bacterium, one of the deadliest known, requires more than 3,000 spores to cause disease.) *Campylobacter,* which thrives at body temperature, incubates for two to eleven days before producing symptoms.

Campylobacter's main virulence factor is probably its ability to invade and cause tissue damage inside the intestine. Initially victims develop a fever, headache, and muscular pain—signs that the bacteria have invaded the body. Typically the body senses proteins on the outer membranes of these bacteria by producing signaling molecules called cytokines. Cytokines affect how immune cells communicate with one another. It is the cytokines, not the bacteria themselves, that induce the initial symptoms. Nausea, abdominal pain, and diarrhea resulting from an inflamed gastrointestinal tract generally follow. As with other food-borne diseases, the inflammation is a result of the host's response to the bacteria destroying one cell after another as it adheres to and multiplies in the intestinal lining.

The vast majority of people stricken with campylobacteriosis recover fully in one to two weeks. However, some develop such complications as reactive arthritis. In rare cases, patients develop Guillain-Barré syndrome, in which peripheral nerves become damaged. When that happens, the victim experiences weakness, numbness, and perhaps tingling, which usually begin in the legs and spread to the arms. As the syndrome worsens, near-complete paralysis can occur. Speech, swallowing, and breathing also may be impaired. Some patients recover completely, although it can take up to a year or longer. Others suffer recurrent bouts of Guillain-Barré syndrome or develop permanent weakness in affected body parts. Of the estimated 1.1 million to 7 million Americans who will be stricken with campylobacteriosis this year, 110 to 1,000 will die.

I. Kaye Wachsmuth, an official in the Office of Public Health and Science at the USDA's Food Safety and Inspection Service, says *Campylobacter*'s special nutritional requirements do not allow it to amplify well in food. As a result, she says, *Campylobacter* infections tend to be sporadic and diffused throughout the population as opposed to some of the more clearly defined outbreaks that have been associated with *E. coli* O157:H7. Because *Campylobacter* is commonly found in stool specimens from people with diarrhea, and because it takes relatively few *Campylobacter* cells to cause illness, government officials consider *Campylobacter* an important public health threat, Wachsmuth says.

Listeria monocytogenes

Another infection that is increasingly in the spotlight is caused by *Listeria monocytogenes*. This hearty bacterium can grow in temperatures ranging from the 37°F inside your refrigerator to the 104°F inside the trunk of your car on a summer day. It can also withstand a relatively wide range of acidity levels, including the pH conditions in milk. Symptoms of listeriosis infection usually begin as fever, muscle aches, and nausea or diarrhea. If the infection spreads to the nervous system, it can cause headache, stiff neck, confusion, loss of balance, or even convulsions. In pregnant women, *Listeria* can cross the placenta and cause premature delivery, listeriosis in the newborn, or stillbirth.

While serious listeriosis infections are infrequent, the disease can be fatal: listeriosis kills about 425 of the 1,850 people who become seriously ill with it each year in the United States, according to the CDC. *Listeria* gained notoriety during an outbreak in California several years ago in which eighty-three people who ate *Listeria*-infected cheese were sickened. Thirty percent of those victims died. In January 1997, the *New England Journal of Medicine* reported another *Listeria* outbreak, this one traced to chocolate milk served at a picnic in Illinois. In that instance, forty-five people had symptoms associated with listeriosis—gastrointestinal distress and fever—and stool cultures from eleven patients yielded the organism.

Listeria monocytogenes is widespread in our environment, existing in soil, water, leafy vegetables, decaying corn and soybeans, raw and treated sewage, effluent from poultry and meat processing plants, and improperly fermented animal feed preserved in silos. *Listeria* in soil or in manure that is used as fertilizer can contaminate vegetables. The bacterium has been found in hot dogs, lunch meats, pâté, spreads, and imported nonpasteurized soft cheeses, specifically feta, Brie, Camembert, blue, and the Mexican-style soft white cheeses queso blanco and queso fresco. Unpasteurized milk and products made from raw milk can also spread the bacterium to humans.

Shigella

As with *Salmonella* and *Campylobacter, Shigella* organisms begin their attack by attaching to and penetrating epithelial cells of the intestinal lining. As they divide and spread to surrounding cells, tissue destruction occurs. This results in shigellosis, also known as bacillary dysentery, which is marked by abdominal pain, cramping, diarrhea, fever, and vomiting, and by blood, pus, or mucus in the stools. In susceptible people, it may take only ten *Shigella* cells to cause symptoms. This low infectious dose makes person-to-person infection a particular problem with this bacterium. As mentioned earlier, some strains of *Shigella* produce a toxin very similar to that produced by *E. coli* O157:H7. Shigellosis symptoms usually occur anywhere from twelve to fifty hours after ingesting the pathogen.

According to public-health officials, an estimated 300,000 annual cases of shigellosis account for less than

10 percent of reported outbreaks of food-borne illness in this country.

Shigella is most commonly found in water polluted with human feces. The disease can also be spread by unsanitary food handling. Foods that have been associated with shigellosis outbreaks include potato, tuna, shrimp, macaroni, and chicken salads; raw vegetables, milk and dairy products, and poultry.

Vibrio vulnificus

This naturally occurring marine bacterium is most commonly found in Gulf of Mexico oysters but may also contaminate clams, crabs, and oysters from the Atlantic Coast as far north as Cape Cod, and from the West Coast, according to the Food and Drug Administration. *Vibrio vulnificus* infections have also been traced to eating fish from brackish lakes in New Mexico and Oklahoma. Because *V. vulnificus* does not result from water pollution, eating oysters from "clean" waters or in reputable restaurants does not provide protection against getting infected by this bacterium, nor does consuming raw oysters with hot sauce or alcohol, according to the FDA. Infections are most prevalent in the summer months because the bacteria proliferate in a warm environment.

Healthy people who ingest *V. vulnificus* may develop gastroenteritis within sixteen hours, although some exposed individuals never develop symptoms. In people with certain underlying medical conditions, including liver disease (particularly alcohol-related liver disease), cancer (especially if the patient is on anticancer drugs or radiation therapy), inflammatory bowel disease, AIDS, diabetes, achlorhydria (reduced stomach acidity), and hemochromatosis (an iron disorder), *V. vulnificus* can cause "primary septicemia," a life-threatening complication that results from the rapid multiplication of bacteria and the presence of their toxic products in the bloodstream. Chills, fever, nausea, vomiting, septic shock, and death can occur in as little as six hours after septicemia symptoms begin. Among people with liver disease, the death rate from *Vibrio* infection exceeds 50 percent. Liver disease can adversely affect the production of antibodies and other factors important in fighting off *Vibrio* and other infections. Also, people with liver disease often have a poor nutritional status, another risk factor for food-borne illness.

Ingesting contaminated shellfish isn't the only way in which people can become infected with *Vibrio*. Swimming in seawater containing *Vibrio* has also been shown to make people sick.

For some reason, alcoholics have a greater risk of *Vibrio vulnificus* infections even if they do not have liver disease. One possible reason is that alcohol makes white blood cells less effective in fighting off the bug. Also, antibodies produced against *Vibrio* may not work as well in alcoholics as they do in nonalcoholics.

Clostridium perfringens

Intense abdominal cramps and diarrhea generally occur eight to twenty-two hours after eating foods that are highly contaminated with the preformed toxin produced by this bacterium. Symptoms usually disappear in twenty-four hours but can persist for one to two weeks in the elderly or the infirm. A few deaths have been reported as a result of dehydration and other complications.

The most common cause of poisoning by *C. perfringens* is failure to promptly refrigerate prepared foods containing meat products or gravy. Small numbers of the organisms are often present after cooking. At room temperature they can multiply to dangerous levels and produce the toxin that causes the disease.

Bacillus cereus

This bug causes two distinct clinical syndromes. The most common is called an emetic syndrome and it is marked by nausea and vomiting that occurs within a few hours of exposure to the preformed toxin that this bug produces in food (typically rice). The other syndrome features a noninflammatory form of diarrhea—very similar to that in the *C. perfringens* picture. Symptoms of *Bacillus cereus* food poisoning (or intoxication) mimic those of *C. perfringens* food poisoning. Watery diarrhea, abdominal cramps, and pain begin within a few hours after eating contaminated food. Nausea may accompany diarrhea, but vomiting is rare. Symptoms generally persist for twenty-four hours.

Many different foods, including meats, milk, vegetables, and fish, have been associated with this diarrheal-type food poisoning. Emetic syndrome from *Bacillus cereus* is usually associated with such starchy foods as rice, potatoes, and pasta, and cheese, but it has also been traced to food mixtures such as sauces, puddings, soups, casseroles, pastries, and salads.

Yersinia enterocolitica

Each year, an estimated 17,000 Americans develop yersiniosis, which is often characterized by fever and by abdominal pain that mimics appendicitis. These symptoms may or may not be accompanied by diarrhea, vomiting, or both. Symptoms generally begin one to two days after exposure to the bacterium. *Y. enterocolitica* causes reactive arthritis in about 2 or 3 percent of cases. Another rare complication is bacteremia. Very few fatalities have been associated with this organism, which has been found in pork, beef, lamb, oysters, fish, and raw milk.

Staphylococcus aureus

Certain strains of *Staphylococcus aureus* can produce enough toxin in food to make you sick very fast, usually in one to six hours. The most common symptoms of staphylococcal food poisoning are nausea, vomiting, retching, watery diarrhea, abdominal cramps, and lightheadedness. In more severe cases, the victim also suffers headaches, muscle cramping, and temporary blood-pressure and pulse-rate changes. Most patients recover in two to three days.

S. aureus (also known as *Staph* infection) is usually spread to food from the hands, coughs, or sneezes of infected individuals. This bug has been isolated from meats, prepared salads, cream sauces, and cream-filled pastries.

Clostridium botulinum

Food-borne botulism is a severe form of food poisoning. It occurs when people ingest foods containing the potent toxin that is formed during the growth of *Clostridium botulinum*. Neurological disease can result from ingestion of just a few nanograms (one nanogram equals one-billionth of a gram) of this preformed toxin. Symptoms usually occur eighteen to thirty-six hours after ingestion of tainted food. Early signs of botulism include profound fatigue, weakness, and vertigo, usually followed by double vision and progressive difficulty in speaking and swallowing. Breathing difficulty, muscle weakness, abdominal bloating, and constipation may also occur. The disease can be fatal if not treated immediately. The recommended treatment for food-borne botulism includes early administration of botulinal antitoxin and intensive supportive care, including mechanical breathing assistance.

Most of the ten to thirty botulism outbreaks reported each year in the United States are traced to inadequately processed, home-canned foods. However, commercially produced foods are also occassionally implicated. The most frequent foods associated with human botulism are sausages, meat products, and canned vegetable and seafood products. Botulism can also result from consuming foil-wrapped baked potatoes held at room temperature.

C. botulinum is ubiquitous in nature. According to the FDA, the organism and its spores have been found in soils; bottom sediments of streams, lakes, and coastal waters; inside the intestinal tracts of fish and mammals, and in the gills and innards of crabs and other shellfish. *C. botulinum* and certain other bacteria and fungi produce spores, thick-walled structures, to protect their cells. These spores are heat-resistant and can survive in foods that are incorrectly or minimally processed. The toxin they contain can be destroyed if heated to 176°F for ten minutes or longer.

Hepatitis A Virus (HAV)

Unlike a bacterium, which may grow and multiply independently on food or in the body, a virus—the smallest known infectious agent—requires living cells in order to replicate. Hepatitis A virus is spread through the feces of infected people and can produce a relatively mild illness in susceptible individuals who ingest fecally contaminated water or foods. For instance, HAV may be transferred from hands to food if an infected individual fails to wash up after using the toilet, or if a caregiver doesn't wash hands thoroughly after changing the diaper of an infected baby. Outbreaks have most frequently been linked to water, shellfish, and salads, although cold cuts and sandwiches, fruits and fruit juices, milk and milk products, vegetables, and iced drinks have also been implicated. Many outbreaks have been blamed on infected workers in food-processing plants and restaurants. A vaccine against hepatitis A is available.

HAV's incubation period, or the amount of time it takes after exposure to a germ for symptoms to emerge, ranges from ten to fifty days, depending on how many infectious particles were consumed. Symptoms usually include a sudden onset of fever, malaise, nausea, appetite loss, and abdominal discomfort, followed in several days by jaundice. Once you have been infected with HAV, you are immune to future infections.

Norwalk viruses

Water—including that from municipal supplies, wells, lakes, swimming pools, and cruise ships—is the most common source of outbreaks from this family of viruses, which also can be spread by infected food handlers. Implicated foods have included raw shellfish, salad ingredients, and cake frosting. An estimated 65 percent of nonbacterial gastroenteritis cases in the United States is attributable to Norwalk and Norwalk-like viruses, with about 181,000 cases occurring annually. Deaths are extremely unusual.

The disease, usually mild and brief, begins one to two days after ingesting contaminated food or water and ends within three days. Common symptoms are vomiting, diarrhea, and abdominal pain. Headache and low-grade fever also may occur.

Scombroid Fish Poisoning

Scombroid fish poisoning is basically an allergic reaction to histamine, a chemical compound that accumulates on certain species of saltwater fish, including tuna and mackerel, if they are not kept sufficiently cold after capture. Histamine is not destroyed by cooking, freezing, smoking, curing, or canning. Symptoms usually begin within an hour, sometimes minutes, of eating histamine-tainted fish. Victims typically experience flushing of the face, arms, and upper torso; severe headache; increased heart rate; stomach cramps with or without diarrhea; facial itching; difficulty swallowing or breathing; and muscle weakness. Most people recover completely in eight to twelve hours. The condition can often be treated effectively with over-the-counter antihistamines. According to seafood specialist Laura Garrido, of the University of Florida Bureau of Seafood and Agriculture, there is no way to tell by appearance, taste, or smell whether a fish you are eating has scombroid poisoning; only a laboratory test can detect the problem.

Ciguatera

Although its occurrence is rare in the United States, ciguatera illness still ranks as one of the top seafood-borne illnesses. It is caused by a natural marine toxin that concentrates in such tropical fish species as amberjacks, moray eels, barracuda, and to a slightly lesser extent, certain snappers and groupers. Ciguatera toxins originate from several algae species common to ciguatera-endemic regions in the lower latitudes. Fish consume the toxin along with the algae. People who consume these fish may develop nausea, cramping, and vomiting followed by headache, flushing, muscle aches and weakness, tingling and numbness in the lips and mouth, and dizziness. More severe cases may include a cold-to-hot sensory reversal in which cold things feel hot to the touch, and vice versa. The onset of symptoms can occur within six hours or less of eating toxic fish. According to the FDA, the occurrence of toxic fish is sporadic, and not all fish of a given species or from a given locality will be toxic. Like histamine, ciguatera toxins cannot be detected by a fish's appearance, taste, or smell, and cannot be destroyed by cooking, freezing, smoking, curing, or canning.

Giardia lamblia

Giardia lamblia is a waterborne protozoan, a single-celled microscopic animal. It causes giardiasis, the most frequent cause of nonbacterial diarrhea in North America. The protozoan, or parasite, has been isolated from domestic animals, such as dogs and cats, and wild animals, such as beavers and bears.

Giardiasis symptoms may include diarrhea within one week of ingestion of the organism's cysts, or sacs containing several protozoa. Normally illness lasts for one to two weeks, but there are cases of chronic infections lasting months to years. The disease mechanism is unknown; some investigators report that the organism produces a toxin, while others are unable to confirm existence of a toxin. It may be that the organism obstructs the absorptive surface of the intestine. Disease can result from ingestion of one or more cysts.

Giardiasis is most frequently associated with the consumption of contaminated water, although at least five outbreaks have been traced to food contamination by infected or infested food handlers. Major outbreaks are associated with contaminated water systems that do not use sand filtration or that have a defect in the filtration system. According to the FDA, the largest reported food-borne outbreak involved twenty-four of thirty-six people who ate *Giardia*-contaminated macaroni salad at a picnic.

Giardiasis is more prevalent in children than in adults, possibly because many individuals seem to have a lasting immunity after infection. This organism is implicated in 25 percent of gastrointestinal disease cases and may be present asymptomatically. It is estimated that 2 percent of the U.S. population has been exposed to *Giardia lamblia.* This disease afflicts many homosexual men, both HIV-positive and HIV-negative individuals, presumably owing to sexual transmission. The disease is also easily spread in daycare centers.

About 40 percent of those diagnosed with giardiasis demonstrate intolerance to sugars, mainly milk sugar, during infection and for up to six months after the infection can no longer be detected. Chronic infections lead to a malabsorption syndrome and severe weight loss. In some immune-deficient individuals, giardiasis may contribute to a shortening of the life span.

Cryptosporidium parvum

Very few people had heard of *Cryptosporidium parvum* until 1993 when more than 400,000 people in Milwaukee, Wisconsin, developed diarrhea after drinking water contaminated with this single-celled parasite. Cryptosporidiosis generally occurs two to ten days after exposure, but not everyone who ingests the organism gets sick. Symptoms include watery diarrhea, headache, abdominal cramps, nausea, vomiting, and low-grade fever. If symptoms are severe, weight loss and dehydration can result. Symptoms generally last one to two weeks.

Cryptosporidiosis is spread by swallowing water or raw or undercooked food that has been contaminated with the stool of an infected person or animal. The disease can also be spread by the classical fecal-oral route in which someone touches a surface containing a microscopic amount of infected fecal matter and then puts his/her hands in his/her mouth. Infections can also occur by swimming in contaminated pools, lakes, or ponds.

Cyclospora cayetanensis

•

This is another one-cell parasite contracted by ingesting water or food that was contaminated with infected stool. The most notable *Cyclospora* outbreak was in 1996, when people in several states and Canada were infected by fresh raspberries imported from Guatemala. Because *Cyclospora* needs days or weeks after being passed in a bowel movement to become infectious, it is unlikely to be passed directly from one person to another.

Cyclospora attacks the bowel and usually causes watery, sometimes explosive diarrhea. Other symptoms may include appetite loss, weight loss, bloating and gas, stomach cramps, nausea, vomiting, muscle aches, low-grade fever, and fatigue. Some people infected with *Cyclospora* have no symptoms. People usually get ill about a week after becoming infected. Untreated, the illness may last a few days to more than a month. Symptoms of cyclosporiasis may wax and wane.

Toxoplasma gondii

Toxoplasma gondii is a protozoan that will probably infect almost any mammal or bird with which it comes into contact. But when it comes time for this parasite to produce eggs, its favorite refuge appears to be a cat's intestinal tract. The disease, toxoplasmosis, can be transmitted to humans through the inadvertent ingestion of these eggs, or oocysts, in cat feces. Another way to become infected is to ingest the organism during another stage of its life cycle in raw or undercooked meat.

Most humans infected with *T. gondii* exhibit no symptoms. However, under certain conditions, toxoplasmosis can cause serious illness, including hepatitis, pneumonia, blindness, and severe neurological problems. These complications are most likely to occur in AIDS patients and others with compromised immune systems.

Toxoplasmosis can also be transmitted through the

placenta to a fetus, resulting in miscarriage, stillbirth, or a newborn with severe mental, and possibly physical, handicaps.

The aforementioned pathogens do not represent all of the known food-borne biological hazards—there are dozens more. They are, however, the most common ones in the United States. Just how much of a threat they are to you depends on who you are.

High-Risk Groups

The vast majority of people who develop food-borne infections recover relatively quickly. After few days of running to the bathroom, the episode is over. Others suffer the consequences of food-borne illness for weeks, months, years, or even a lifetime. As the *Salmonella* scenario at the beginning of this chapter suggests, the healthiest among us can be stricken, and many experts are convinced that these infections occur far more often than most people realize. Epidemiologist Michael Osterholm estimates that Minnesota's 4.4 million residents suffered 6.1 million cases of diarrheal illness over a twelve-month period spanning 1996 and 1997. While not all of these cases were related to food-borne pathogens, he says, many were.

"Anytime an individual has diarrhea or an upset stomach, the first thing that should come to mind is food-borne illness," maintains food scientist Christine Bruhn.

How sick you get and how quickly you recover are

partly determined by the number and virulence of the pathogens you ingest. But the condition of your immune system and your overall current health status are other important determinants.

Listed below are groups that should consider themselves at high risk for food-borne infections. Together, these high-risk groups represent about 20 percent of the U.S. population. Compared to the general population, high-risk individuals often require a smaller dose of infectious agents to get sick, and their symptoms tend to be longer lasting and more debilitating. People in high-risk categories are also more likely to die or suffer chronic health problems as a result of a food-borne infection. Risk status changes throughout life; for example, advanced age and certain age-associated diseases are risk factors. Your doctor or nutritionist can help you decide whether you need to avoid high-risk foods and take extra steps to prevent food-borne diseases.

The high-risk populations are:

- The very young
- The very old
- People with AIDS, HIV, or other immune disorders
- Antacid users
- People on antibiotic medications
- Pregnant women
- Cancer chemotherapy patients
- People with liver disease
- Organ transplant recipients
- People who have undergone gastric surgery

- Long-time steroid users (for such conditions as asthma or arthritis)
- The malnourished
- The mentally ill and mentally challenged

Diabetics are considered at high risk for *Vibrio* infections and should therefore avoid raw shellfish. They are at "intermediate" risk for most other food-borne diseases, which means they are slightly more susceptible than the general population. If you have any chronic medical condition not mentioned above, ask your doctor if you need to avoid risky foods, such as raw shellfish.

The very young. The best way to build up your immune system is to expose it to a variety of germs. The immune system then "remembers" how it fought off each germ. The next time that germ or one like it tries to invade your body, your immune system is ready to pounce. It usually destroys the enemy organism before symptoms emerge. Babies' and young children's immune systems have encountered fewer germs than adults', and this is presumably why youngsters are more vulnerable to food-borne illnesses and other infectious diseases.

A child's environment plays an important role in how the immune system matures. For example, researchers have found that people living in rural areas have higher-than-average levels of Shiga toxin antibodies in their blood. It is difficult to know whether people reared in rural areas are protected against Shiga-toxin-producing *E. coli* O157:H7, however, because it is unethical to deliberately expose humans to a potentially fatal infection.

How long it takes for the immune system to strengthen depends largely on how many germs a person is exposed to early in life. For example, many people get exposed to the waterborne protozoan *Cryptosporidium* when they are young. The infection typically produces a mild illness but usually confers immunity to further attack.

This sort of acquired immunity is determined in large part by where a child grows up. For example, a child reared on a farm spends a lot of time around animals and is therefore likely to build up immunity to many more food-borne pathogens than a child growing up in a place like Boston or Los Angeles. Likewise, a baby born in a developing country may have more naturally acquired immunity than a typical American baby. For example, in 1990, a gastroenterologist working in a hospital in India noted that even when *Campylobacter* was isolated from the stool of a patient with diarrhea, it was necessary to search for other causes of diarrhea, because although *Campylobacter* is well known in many parts of the world for its gastrointestinal symptoms, in India it is so pervasive in food that virtually everyone living there has had exposure to it and has therefore developed immunity to it. When diarrhea occurs in that country, it is almost always caused by something other than *Campylobacter*.

In Korea there may be some naturally acquired immunity for certain strains of *Salmonella*. In 1995 a routine stool test performed on a Korean-born boy named Ethan turned up positive for *Salmonella* the day after he was adopted by a New Jersey family. His parents were perplexed because Ethan, who was four months old at the time of his adoption, appeared healthy and had no symptoms of salmonellosis. Ethan's mother was told to send stool samples to the local health department for testing once a month. After six months, the results finally came back negative.

For three-year-old Dylan, who was born in Pennsylvania, exposure to *Salmonella* produced such brutal bloody diarrhea, appetite loss, and abdominal pain that he had to be hospitalized for three days. Dylan's mother, Jeanne, strongly suspected that Dylan had picked up *Salmonella* during a family party at which he was the only child. Smoked fish, fish salad, egg salad, and tuna salad were served buffet style and had been left at room temperature for at least two hours. Everyone ate the same food, but Dylan was the only one who became ill.

In general, newborns may be protected from certain food-borne pathogens because maternal antibodies they received in the womb continue to circulate in their blood for a limited amount of time. If a baby is breast fed, this immunity is usually augmented. Weaned babies, toddlers, and preschoolers may be the most vulnerable groups at their end of the age spectrum.

The very old. Most Americans in their seventies are in generally good health, but as they reach eighty years of age and beyond, certain physiological and circumstantial changes put them at increased risk for food-borne diseases. For example, the stomach produces less acid as people age, and immune system function slowly declines. Also, when an elderly person gets diarrhea and has to run to the bathroom every twenty minutes, the illness is probably going to weaken him more quickly than it would a thirty- or forty-year-old. This is especially true for senior citizens who suffer from energy-sapping chronic medical conditions, such as emphysema or heart failure. All sorts of insults to the body, including food-borne pathogens, are more difficult to cope with when you are elderly.

Additionally, any decline in cognitive functioning can put the elderly at risk. Alzheimer's disease and other age-related dementias will, by definition, impair a person's ability to handle food safely and practice good personal hygiene.

Immune disorders. Anyone with AIDS, lymphoma, leukemia, Hodgkin's disease, or another disease that weakens the body's immune response has difficulty fighting off food-borne infections and should therefore avoid risky foods and take other precautions. To a lesser extent, those with HIV also have an elevated risk because their immune systems are under attack.

Once someone with HIV is diagnosed with AIDS, studies have shown that the risk for *Salmonella* infection rises 20- to 100-fold. If a person with AIDS contracts salmonellosis, there is a greater-than-average risk it will develop into septicemia, a life-threatening condition. Another study found that AIDS patients had a 35-fold increased risk for *Campylobacter* infection. In the San Francisco area, investigators found a 280-fold increase in the incidence of listeriosis among AIDS patients. *Cryptosporidium,* meanwhile, has been found in an estimated 10 to 20 percent of cases of AIDS-associated diarrhea. In the February 1998 issue of the *American Journal of Epidemiology*, researchers at Columbia University School of Public Health in New York advised HIV-infected individuals to boil their tap water before drinking it to prevent the small risk of cryptosporidiosis. Although rare, this disease can be life threatening in people with suppressed immune function.

Antacid users. Pepcid AC, Zantac, and similar medications reduce the acidity of the stomach lining, lowering the likelihood of a person developing a peptic ulcer. Unfortunately, a reduction in stomach acidity raises an individual's risk of contracting a food-borne infection. Pathogens that would normally die in the stomach may survive in the less acidic environment.

Antibiotics. The most common antibiotics being prescribed today don't necessarily discriminate between "good" and "bad" bacteria in your body. Your digestive tract, through which the antibiotic enters your system, normally contains billions upon billions of good bacteria, or flora, that aid in digestion. Normally your flora compete with any pathogenic bacteria you may ingest, reducing your risk of becoming ill. When an antibiotic kills off all or most of this good bacteria, any germs you ingest are more likely to find a home in your intestine, where they can colonize and do harm. Eating yogurt that contains live cultures may help replace some of the flora temporarily. But it probably takes about two weeks after finishing your antibiotic prescription before the bacteria in your gut return to normal levels. Until then, consider yourself more vulnerable than usual to food-borne infections.

Pregnancy. The USDA strongly advises expectant mothers and the people preparing their food to be especially diligent when following safe food-handling recommendations. Any disease a pregnant woman contracts may adversely affect her pregnancy or unborn child. The classic food-borne threat during pregnancy is listeriosis, particularly during the third trimester when certain components of a woman's immune-response system become less effective. *Listeria* can take up residence in the placenta, damage it, and reduce the fetus's blood supply. This can lead to premature labor, miscarriage, and other problems. The organism can also invade a pregnant woman's bloodstream, passing the disease to her fetus.

Chemotherapy. Cancer-fighting drugs appear to raise one's risk for food-borne illness on several different fronts. Most chemotherapy drugs indiscriminately attack rapidly dividing cells, including those in the bone marrow where infection-fighting white blood cells are made. This temporarily hampers the body's ability to fight off all kinds of germs, including food-borne pathogens. If the chemotherapy cocktail includes a steroid medication, which many do, the patient's immune response is further reduced. Radiation treatments can also temporarily impair the immune system.

Many cancer chemotherapeutic agents have antimicrobial properties that reduce colonies of beneficial (digestion-aiding) bacteria in the stomach and intestine.

With less competition for nutritional resources, any pathogenic bacteria that enter the intestinal tract have a better opportunity to hunker down and colonize.

Chemotherapy also theoretically deteriorates the intestinal lining by attacking the rapidly dividing epithelial cells that line the digestive tract. (This is also why chemotherapy often triggers diarrhea.) It is possible that disease-causing bacteria may have a slightly easier time invading the intestinal wall after some of the epithelial cells have been destroyed by the drugs.

Liver disease. The liver is the body's main detoxification organ. If it is not functioning at peak efficiency, it cannot fully protect you against food-borne pathogens or the toxins that some of these bugs produce. Alcoholics with cirrhosis of the liver are particularly vulnerable to *Vibrio vulnificus,* the bacterium most often associated with raw oysters. Nonsymptomatic liver disease has been detected in people who drink two or three alcoholic beverages a day. The FDA says the risk of death from *V. vulnificus* infection is almost 200 times greater in people with liver disease than in those without liver disease.

Organ recipients. According to the United Network for Organ Sharing, the number of organ transplants performed in the United States has been rising each year. More than 20,300 kidneys, hearts, lungs, and other organs were transplanted in 1996, the most recent year for which data are available. In order to prevent organ rejection, almost all recipients must take antirejection drugs, which suppress the immune system, for the rest of their lives. These medications leave these individuals more vulnerable than the general population to the ravages of food-borne infections, as well as other infectious diseases.

Gastric surgery patients. People who lose all or part of their stomach due to disease, or as a treatment for obesity, have less gastric acid available to destroy food-borne toxins or pathogens before they reach the intestine.

Long-time steroid users. People with severe asthma or chronic arthritis may be given steroids to help manage their disease. Unfortunately, steroids can impair the body's response to infection, thereby allowing mild infections that are normally fought off to take hold and potentially cause major problems.

Malnourishment. People, particularly children, who don't get adequate calories or nutrients are less able to fight off food-borne infections and other infectious diseases.

Mental problems. People with serious psychiatric disorders, such as schizophrenia, and people who are developmentally disabled are at increased risk of acquiring and spreading food-borne disease if their personal hygiene habits are poor. They do not, however, have a higher risk of complications if they are in good physical health.

What to Do if You Are in a High-Risk Group

If you are in a high-risk group, avoid eating raw or undercooked fish, meat, and poultry; drink only pasteurized juices and milk; and know the source of the water you drink. Fastidiously following the food-safety guidelines in Chapters 9 and 10 will help reduce your risk of infection. Don't hesitate to go that extra mile to avoid possible exposure to food-borne pathogens, even if it means walking out of a restaurant that refuses to show you its latest sanitation report or using separate cutting boards for meat, poultry, sandwiches, and produce at home. Taking such precautions may seem excessive, but they can save your life.

If you do develop diarrhea, vomiting, or related symptoms, call your doctor immediately. As will be discussed in the next chapter, accurate diagnosis and prompt, aggressive treatment can often avert disaster.

3

DIAGNOSIS AND TREATMENT OF FOOD-BORNE INFECTIONS

You are feeling nauseated. You have stomach pain and diarrhea. Perhaps your head also hurts, and you are starting to develop a fever. Is it just a "twenty-four-hour bug," or is it something you ate? Distinguishing between the two is not always easy, especially during the initial stages of a food-borne disease.

Often the first sign of a food-borne illness is nausea, a feeling of being sick to your stomach, which may or may not manifest itself in vomiting. But nausea can also stem from a host of other causes, including emotional upset, fever owing to an infection elsewhere in the body, or a medication that is stimulating the vomit center in the brain—to name a few. When the trigger is a food-borne pathogen, the next symptom is usually abdominal pain or cramping as the pathogen and, in some cases, its toxin amplify in the small intestine. The severity of abdominal discomfort varies widely from person to person; some people feel a little bloated, others double over in agony, and some have no cramping at all. Two siblings who receive the same dose of the same pathogen can fall on different points of the pain spectrum.

If you have ingested *Salmonella, Campylobacter,* or another food-borne bug, the third likely symptom is some kind of diarrhea. Again, there are many possible causes of diarrhea, some infectious, some not. When it is caused by a food-borne infection, diarrhea is typically loose or watery, and it may or may not be bloody, explosive, or contain mucus. Of course, there are always exceptions to the rules. With some pathogens, such as *Bacillus cereus,* which causes food poisoning from a preformed toxin, symptoms may stop at the stomach. The episode passes after a few hours of nausea and vomiting.

Almost anything that upsets the body's homeostasis—its ability to maintain a constant internal environment—can cause gastrointestinal (GI) upset. Fever from any source can certainly do that, and infections—food-borne and nonfood-borne—are a major cause of fever.

Masters of Disguise

Adding to the confusion of distinguishing between food-borne and nonfood-borne infections are noninfectious diseases that can masquerade as food-borne illness. Among the most common are:

- **Inflammatory bowel disease.** There are two main types of inflammatory bowel disease: ulcerative colitis and Crohn's disease, both of which tend to begin in early adulthood. Ulcerative colitis refers to open sores and chronic inflammation in the lining of the colon (the large intestine) and the rectum. Only the rectum may be involved in the early stage of the disease. Crohn's disease refers to a chronic inflammation that can strike any part of the digestive tract but most commonly the terminal ileum—the end of the small intestine where it meets the large intestine. Both Crohn's disease and ulcerative colitis can seem very much like food poisoning, with severe abdominal pain, intestinal spasms, and bloody, mucus-laced diarrhea. A thorough family history, a physical exam, and sometimes a colonoscopy or sigmoidoscopy to visually inspect the lower intestine or to obtain a biopsy are used to diagnose these diseases.
- **Irritable bowel syndrome.** Sometimes called a spastic colon, irritable bowel syndrome is a widespread, predominantly adult malady frequently triggered by one's emotional state, particularly by extreme anxiety or stress. Prior to

a job interview or a major exam, for example, the patient may have three or four watery stools, sometimes accompanied by abdominal cramps. Some cases of irritable bowel syndrome manifest in constipation.

If your diarrhea is occurring during a time of stress in the absence of vomiting and fever, it is unlikely that you have a food-borne infection. Also, infectious diarrhea and inflammatory bowel disease will often wake you at night because of the need to go to the bathroom; irritable bowel syndrome is much less likely to do that.

- **Ischemic colitis.** This disorder is found primarily in the elderly. Like food-borne infections, ischemic colitis typically causes abdominal pain and bloody diarrhea, and it may also spawn a fever. It is caused by vascular disease in the intestine, and it is analogous to coronary artery disease in the heart. Basically the blood vessels feeding the intestine become narrowed or blocked, reducing blood flow to that area of the digestive tract. A clinical diagnosis of ischemic colitis is usually confirmed when surgery is performed to correct the problem.
- **Diverticular disease.** Another affliction associated with advanced age, this condition is marked by the formation of outpouchings, or pockets, in the normally smooth intestinal wall. The disease usually affects the lower end of the colon. If fecal matter gets lodged in one or more of these pouches, the area can get inflamed, causing pain, discomfort, and sometimes bleeding

that shows up in the stools. Diverticular disease can be diagnosed with a specialized X ray or a colonoscopy.

- **Intussusception.** This condition, which is most often seen in infants and young children, occurs when the bowel telescopes in on itself, as though a pencil were pushing in on the end of a frankfurter-shaped party balloon. The patient experiences severe abdominal pain (colic in infants) often accompanied by vomiting and by blood and mucus in the feces. The best way to make this diagnosis is by administering a barium enema and then scanning the area. This diagnostic procedure is often therapeutic as well, because the act of passing the barium actually causes the bowel to return to its normal configuration.
- **Lactose intolerance.** An inability to digest lactose, a sugar found in milk, is usually caused by a deficiency of the enzyme lactase in the small intestine. This condition may cause abdominal cramping but it is not associated with fever or other systemic symptoms, such as body aches and headaches. Lactose intolerance can be diagnosed with the aid of a "hydrogen breath test" machine, which is very sensitive and specific. The patient drinks a lactose beverage then breathes into the machine, which periodically measures hydrogen in the breath. If lactose is absorbed as it should be in the relatively sterile environment of the upper intestine, then hydrogen levels in the breath will

> measure less than 10 parts per million. In cases of lactose intolerance, lactose reaches the more distant small bowel, where bacteria break it down, creating hydrogen as a by-product. This hydrogen will diffuse through the gut wall, get into the bloodstream, and ultimately be excreted by the lungs at levels greater than 10 parts per million.

Physicians may take steps to rule out some of the aforementioned maladies even as they consider a diagnosis of food-borne disease.

One diagnosis you are unlikely to get from a qualified physician is "stomach flu." Lay people may use this phrase to describe any kind of gastrointestinal upset, just as they label anything that causes fever, aches, and pains as the "flu." As a medical diagnosis, stomach flu is a misnomer. The term "flu" is short for influenza, a viral infection of the airway passages. A more precise generic description of most food-borne maladies is "gastroenteritis," which refers to inflammation anywhere in the digestive system. There are many possible causes of gastroenteritis, but when the cause is food-borne, symptoms generally work their way from the top to the bottom of the gastrointestinal tract. That is to say the episode begins with nausea and vomiting, progresses to abdominal cramps, and ends with diarrhea, although in some cases it all seems to start at the same time.

When to Seek Medical Attention

If everyone in America who contracts a food-borne disease each year ran to their doctors' offices, the medical system (or at least the health insurance industry) would probably collapse. Fortunately the vast majority of sufferers recover within a few days without medical intervention. Many never realize that they even had a food-borne infection.

Certainly anyone at high risk for complications—including the very young, the very old, those with weakened immune systems, pregnant women, and people with liver disease—ought to consult a physician if gastrointestinal symptoms are severe. Occasionally people who don't fall into a high-risk group can also develop serious complications from a food-borne infection and thus may also benefit from prompt medical attention.

The next time you or a loved one experiences gastroenteritis (nausea with or without vomiting, some degree of abdominal pain, plus diarrhea), ask the following questions. If you answer yes to any of them, or if you are in doubt about how to answer them, you should consult your family physician:

1) *Have the symptoms lasted for more than two days?*
2) *Is there a fever?*
3) *Is the diarrhea very watery or explosive?*
4) *Are there signs of dehydration?* If enough fluid is lost through diarrhea or vomiting, it can affect the body's salt balance. Children can become dehydrated more quickly than adults. The main

symptoms of dehydration are intense thirst, dry lips and tongue, increased heart and breathing rates, weakness, and dizziness.

5) *Is blood mixed in with the stools?* Making this determination can be tricky. If the water in the toilet bowl is bloody, but blood is not interspersed throughout the feces, the cause could be hemorrhoids or an anal fissure. Blood laced into the feces indicates that the problem is originating higher up in the intestinal tract and may possibly stem from a food-borne infection. See a doctor if you are not sure. Any time a child has bloody diarrhea seek medical attention immediately and insist on a stool test for *E. coli* and other common food-borne pathogens.
6) *Is pus or mucus (whitish-gray in color) mixed in with the stools?*
7) *Have bowel movements been unusual in frequency or consistency for more than two days?*
8) *Do symptoms include shivers or chills?*
9) *Has anyone else in your family been sick with similar symptoms?*

Don't let an absence of fever lull you into a false sense of safety if you are experiencing severe GI symptoms. The rare but potentially fatal complication of *E. coli* O157:H7, hemolytic uremic syndrome (HUS), doesn't usually generate a high fever, although a low-level fever is quite common.

What Have You Eaten Lately?

If your symptoms seem to point toward a food-borne infection, it can be helpful to review what you have eaten over the last few days. Be careful not to focus solely on yesterday's dinner or this morning's breakfast. If you think you have food poisoning, it is human nature to think back to your last meal. Wondering about that hamburger you ate for lunch, you may forget that four days ago you were at a wedding and had potato salad along with fifty other people who are now scattered around the country and may be having the same symptoms as you. *Campylobacter* may incubate for up to ten days in a human before symptoms occur; *Salmonella* and *E. coli* both incubate for two or three days; bacteria such as *Clostridium botulinum*, which produce a lot of preformed toxin, can make you ill within hours.

Here are some specific questions to ask yourself as you try to figure out how you may have contracted a food-borne infection:

1) *Have any of your recent dining experiences taken place in an out-of-the-ordinary setting, such as a picnic, party, or wedding reception?* If so, it may help to call one or two other people who were there to learn if they have symptoms similar to yours.
2) *Have you eaten raw seafood in the last week or so?*
3) *Have you eaten any other risky foods lately, such as rare hamburger, eggs cooked sunny-side up, salad-bar items, or insufficiently warmed leftovers?*

4) *Have you traveled recently, especially outside the country?*
5) *Have there been any recent power outages in your home that could have allowed food in your freezer or refrigerator to warm up enough to foster bacterial growth?*
6) *Have you recently dined at a restaurant or eaten take-out food?*

How Doctors Diagnose Food-borne Illnesses

The doctor's first step is to take a history of your symptoms: what they are, when they started, how they have progressed, how much fever and pain you have. You may also be examined physically for any abdominal tenderness or masses that might suggest something other than a food-borne disease. If one or more family members have had similar symptoms, the doctor will suspect an infectious agent as opposed to diverticulitis or another noninfectious disease.

Tests. The doctor may ask you to provide a stool specimen to be tested for the presence of pathogenic bacteria. Bacteria can be isolated from a stool specimen in a lab, and most labs are equipped to test for the most common food-borne pathogenic bacteria.

There are several laboratory tests that can determine whether a patient has been infected with *E. coli* O157:H7. They include the MacConkey-Sorbitol test, which delivers results in twenty-four to forty-eight hours; and Meridian's Premier EHEC EIA, which works on both stools and meat, and can deliver results in two to sixteen hours, depending on whether a stool culture is required. A third test, made by Alexon-Trend, has recently received FDA approval. The MacConkey-Sorbitol test looks for the bacteria directly; the other two, which are more costly, probe for the Shiga toxin itself and have the advantage of screening for other Shiga-toxin-producing bacteria that have been incriminated in human disease.

It is extremely difficult to isolate a virus from a stool sample, but viral symptoms are seldom as severe as those caused by bacteria. Identifying protozoa, such as *Cryptosporidium,* requires special laboratory tests that are not routinely done. Because these parasitic infections can be difficult to diagnose, you may need to provide a series of stool specimens over several days.

If your diarrhea is frequent, you may be able to produce a specimen during your doctor's appointment. Otherwise the doctor can give you a sterile cup or stool-sampling kit to take home and return to the office or to a reference laboratory. It generally takes at least twenty-four hours to obtain the results of a stool test. By then symptoms may have cleared up on their own, or antibiotic treatment may already have been started.

A stool test is not always necessary for effective treatment, but it can confirm a clinical diagnosis and signal a possible outbreak. Also, if a laboratory can identify the pathogen that is making you sick, the doctor may be better able to prescribe a specific and appropriate antibiotic. Physicians are being urged to prescribe antibiotics more judiciously these days in light of multidrug-resistant strains of bacteria, food-borne and otherwise.

If the physician reports your condition to local health authorities, the health department may wish to obtain or confirm a laboratory test and, in some instances, request periodic stool specimens from you until the infection passes. Laboratory confirmation of a food-borne disease can also help prevent the spread of the pathogen from person to person because it sharpens awareness in the infected individuals and the people who come in contact with them. Infected people will shed pathogens in their stools.

Good hygiene is always important but it takes on added meaning when infectious microorganisms are being excreted, especially if one of your family members is in diapers. If your baby or toddler is diagnosed with a food-borne infection, be extra careful when changing soiled diapers. Use disposable diapers and consider wearing a fresh pair of plastic or latex gloves for each diaper change. After the change, place the gloves along with the soiled diaper into a plastic bag and knot the bag before disposal. Regardless of whether you wear gloves, be sure to wash your hands thoroughly with soap and warm water for at least twenty seconds after changing a diaper. The changing table should be wiped down with a disinfectant and paper towels. Notify any daycare providers to take similar precautions, even if your child no longer has symptoms.

To Treat or Not to Treat?

As alluded to earlier, most food-borne illneses are what doctors call "self-limiting"—they go away on their own without treatment. Resting and drinking plenty of fluids is all that is needed in the vast majority of cases. Because they can become dehydrated more easily, babies and young children should be given Pedialyte or a similar product at the first sign of diarrhea. If you have mild diarrhea that is not incapacitating, and there is no blood in your stools, and no fever, shivers, or chills, you have the option of taking an over-the-counter drug such as Imodium or Pepto-Bismol to reduce diarrhea and quell queasiness. Be aware, however, that using these remedies for symptomatic relief of food-borne illness is controversial. On the one hand, these preparations will make you more comfortable. On the other hand, they will suppress diarrhea that normally helps express the pathogens out of your system. If you artificially slow down that process, the bugs are going to linger longer inside your intestine.

Before your doctor prescribes an antibiotic for a food-borne bacterial disease, ask for a stool test first to rule out Shiga-toxin-producing *E. coli* such as O157:H7. Some recent laboratory data suggest that it may be potentially dangerous to treat these bugs with antibiotics for two reasons: 1) the antibiotic will kill the bacteria in the intestine, and in the process, a lot more toxin will be released as the bacteria die and fall apart; and 2) certain antibiotics may actually increase the bacteria's toxin output by turning on its Shiga-toxin genes. This does not happen with high-dose antibiotics, but in test-tube experiments, intermediate dosages appear to stress the bacteria in such a way that they crank up their toxin production by ten- or even fifty-fold. If this happens in the human body, the patient may be at an increased risk of developing complications such as kidney failure. Unfortunately, there are no reliable animal studies to back up these findings. From the physician's standpoint, whether to treat *E. coli* O157:H7 infections with antibiotics is a hard call to make.

Once a diagnosis of infection by a Shiga-toxin-producing strain of *E. coli* has been ruled out, antibiotics are indicated only if there is evidence of severe illness: high, prolonged fever with chills; prolonged bloody diarrhea; or severe abdominal pain lasting forty-eight hours or more. Such symptoms suggest that the bacteria may have gotten into the bloodstream, and taking antibiotics at this point will likely shorten the course of the disease and may prevent complications, such as Reiter's syndrome.

Another time to use antibiotics in treating a food-borne illness is when the patient is a professional food handler who is in a "prolonged carriage state." This means that he or she feels fine physically but continues to test positive for a food-borne illness, such as salmonellosis. Under these circumstances, a chef, caterer, cafeteria worker, or employee of a restaurant or food-production plant would not be allowed to return to work unless given the antibiotics, because the risk of contaminating others through food would be too great.

Of course, some people are allergic to certain antibiotics and should avoid them, and some antibiotics are not appropriate for children.

Norwalk and Norwalk-like viruses, a prevalent cause of gastrointestinal upset, usually produce mild gastrointestinal symptoms that disappear in a couple of days without medical treatment. Fluid replacement is the only therapy needed.

Food-borne parasites, such as *Toxoplasma gondii,* are generally treatable with antibiotics.

Treating complications of food-borne disease. One of the most dangerous complications associated with food-borne infections is hemolytic uremic syndrome (HUS), a disorder that can lead to kidney failure, blood clots in the kidneys and brain, heart failure, respiratory distress, intestinal tears, pancreatic damage, and death. Caused by Shiga toxin, HUS occurs in up to one in twenty cases of pathogenic *E. coli* infections, and the majority of victims are children. HUS survivors often are left with permanent physical impairment, or they face an elevated risk of developing kidney failure or diabetes.

Several possible antidotes to *E. coli* infection are under investigation, but there is no approved medical treatment for HUS. So the best available therapy is supportive therapy—putting the patient in the hospital where electrolytes (salt and fluid balance), kidney output, blood pressure, and other vital functions can be closely monitored. If kidney failure develops, the patient can be placed on dialysis, and blood or platelet transfusions can be given if needed. Some HUS patients require medication for high blood pressure owing to kidney malfunction. In about one of every twenty HUS cases, the patient dies despite aggressive therapy.

Scientists at the National Institutes of Health are developing a vaccine they hope will provide immunity to *E. coli* O157:H7 infection. The NIH scientists took molecules from the outer coats of *E. coli* O157:H7 bacteria and attached them to a harmless protein that has nothing to do with *E. coli*. When this substance was injected into eighty-seven volunteers, almost all developed antibodies within one week, with no significant adverse effects. More time is needed to determine whether the vaccine actually protects against *E. coli* O157:H7 infection. Even if the vaccine works, there is a major drawback to this bacteria-specific approach: What if millions of dollars are spent vaccinating children against *E. coli* O157:H7, but then along comes *E. coli* O111:NM, which has exactly the same capacity to make Shiga toxin and cause disease and death? The *E. coli* O157:H7 vaccine would be powerless against *E. coli* O111 and other Shiga-toxin-producing bugs that might emerge in the future.

To combat Shiga toxin directly, researchers at New England Medical Center are trying to develop a Shiga-toxin vaccine that can potentially be used in two different ways. The first is called active immunization and is preventive: children (or adults) could be vaccinated against Shiga toxin just as they are vaccinated against tetanus. The Shiga vaccine would prompt the body into developing antibodies that would kick in if the person were exposed to Shiga toxin later in life. The second approach, called passive immunization, is designed for people who have already absorbed Shiga toxin. These patients would receive infusions of anti-Shiga-toxin antibodies that have been raised in some other healthy human donors and pooled for this purpose. Passive immunization works for tetanus, and scientists hope it will also work for Shiga intoxication during a crisis situation. Unfortunately, passive immunization as a treatment for HUS cannot be tested on animals because animals do not get HUS. At this writing, passive and active Shiga-toxin immunization are at least two years away from human clinical trials.

There is at least one other investigative therapy for Shiga intoxication—a compound called Synsorb-Pk, which is being tested clinically in the United States, Canada, Japan, and Argentina. Synsorb-Pk consists of inert silica particles similar to sand but ground up to the consistency of flour. Scientists know which very specific sugars from the surface of human cells the Shiga toxin normally binds to. Investigators have created synthetic versions of these sugars and attached them to the silica particles. When patients drink Synsorb-Pk mixed with water or juice, millions of these coated particles enter the digestive tract and bind to the Shiga-toxin molecules. The idea is to soak up all of the toxin directly from the intestine to prevent its absorption into the blood, where it could travel to the kidneys and trigger HUS.

Synsorb-Pk has two potential drawbacks: 1) the diagnosis of *E. coli* infection must be made very early, before the toxin enters the bloodstream; and 2) no one is sure exactly what percentage of the toxin must be mopped up in order to prevent complications, even if treatment begins on time. Data suggest that Synsorb must bind to 90 to 100 percent of the toxin molecules in order to be fully effective. The upside is that once the toxin molecules stick to the Synsorb, they remain stuck until the particles are passed out of the body during a bowel movement. If Synsorb is approved, it could make sense to administer it in conjunction with an antibiotic. As the bacteria die and release their toxin, the Synsorb would be right there to absorb it.

Pain from reactive arthritis, a complication associated with several food-borne pathogens, can be eased with nonsteroidal anti-inflammatory drugs such as ibuprofen. There is no known cure for Guillain-Barré syndrome, the rare neurological complication of campylobacteriosis. The patient is usually put in a hospital so that the condition can be monitored and breathing assistance rendered if necessary. The syndrome usually clears up on its own after several months, although some patients are left with permanent weakness or episodic flare-ups.

Botulism, one of the most dangerous food-borne diseases, can be diagnosed with a blood or stool test, but these tests take several days to yield results, and a clinical diagnosis is usually made first. Botulinum toxin can cause paralysis that progresses symmetrically downward, usually starting with the eyes and face, to the throat, chest, arms, and legs. When the diaphragm and chest muscles become fully involved, breathing is halted and the victim dies from asphyxia. As soon as possible after diagnosis, the patient should be given botulinal antitoxin, which is available from the CDC, plus intensive supportive care, including the use of a respirator, if necessary.

Surgery. On very rare occasions, a severe case of food-borne disease can result in a perforation, or tearing, of the intestine, which can be repaired surgically. Some patients develop profound hemorrhagic colitis—literally bleeding to death from the colon—and removing the colon surgically can be lifesaving. This, too, is an extremely rare complication.

Misdiagnosis

Food-borne infections are occasionally misdiagnosed, and there are documented cases in which people had unnecessary abdominal surgery—appendectomies and more extensive procedures—when their problem turned out to be food-borne disease.

According to Safe Tables Our Priority (STOP), a consumer advocacy group, a physician may misdiagnose frequent, watery diarrhea and severe stomach cramps as the flu when the symptoms are actually early signs of *E. coli* infection. In a recent outbreak traced to contaminated raw juice, two children were turned away from emergency rooms because they were deemed not sick enough for emergency treatment. STOP further asserts that efforts to lower health-care costs "conspire against diagnosis" by discouraging physicians from ordering the full panel of stool and blood tests for food-borne bacteria.

You may avoid a misdiagnosis and dangerous complications by following the guidelines laid out at the beginning of this chapter and by insisting on laboratory tests for specific food-borne pathogens, regardless of whether antibiotic therapy or surgery is recommended.

Educating Your Doctor

While primary-care physicians successfully treat the vast majority of food-borne illnesses in this country, they may not be fully aware of all of the intricacies in this rapidly advancing area of medicine. Scientific interest in food-borne disease is intense, and new findings are being reported almost every day.

By the time you finish reading this book, there is a decent chance that you will know more about food-borne diseases than your doctor does. Most doctors hate to admit that they do not know everything, and your knowledge will probably make them uncomfortable. Imagine the look on your doctor's face if he or she prescribes an antibiotic and you say: "Can you be sure my illness is not from Shiga-toxin-producing *E. coli,* and will the antibiotic make things worse?" Or imagine a doctor ruling out *E. coli* O157:H7 through a stool test and telling you not to worry about your son or daughter getting HUS. Then you say, "Have you ruled out the other Shiga-toxin-producing *E. coli*? Why not test the stool for Shiga toxin directly?" The doctor may have no idea what you are talking about; your queries may even prompt him or her to search the medical literature.

As you point out these issues to your doctor, try to maintain a gentle, respectful demeanor; do not be an alarmist. An astute doctor will realize that you are an informed patient and will probably go that extra mile and order the test, if you request it. If the doctor refuses, and complications do set in, his/her legal liability would be inestimable. (If the doctor refuses to order the test, you can always consult another doctor.)

The Big Picture

Reporting a suspected or confirmed food-borne infection to the health department won't affect your recovery, but it can potentially protect others from getting the same illness. Your information may help identify an outbreak in a restaurant or other public place so that steps can be taken to prevent future outbreaks. Cooperating with health officials enables you to bring something positive out of your negative experience with food-borne disease; your actions may even save a life.

Laws vary from state to state regarding the reporting of infectious diseases. These laws primarily cover health-care providers, but local and state health officials also appreciate hearing from the public about food-borne infections, says Jesse Greenblatt, M.D., M.P.H., state epidemiologist for New Hampshire and a consultant to the Council of State and Territory Epidemiologists. In fact, he says, a significant proportion of food-borne illnesses are reported by nonphysicians. You do not have to know which pathogen you ingested, and you do not need a positive stool test in order to notify the health department. An intuition that your gastrointestinal symptoms are food-borne is reason enough to call.

If you suspect that you have a food-borne illness, you can report it to your local or state health department; these offices compare notes and often team up to investigate possible outbreaks. Health departments also act as clearinghouses; families living in different parts of the same city may have gotten infected at the same restaurant, and no one would find out unless more than one party reported their symptoms.

When a public-health specialist agrees that an illness is possibly food-borne, the caller is generally asked to provide a three-day food history. Health officials will also want to know, among other things, whether anyone else in the family is sick. Most of the time, the diarrhea is mild enough to suspect a viral infection. If there is no apparent cause, the case is usually closed. If symptoms are severe, and more than one person is reporting them during the same general time frame, it can tip off health officials to the possibility of a more serious bacterial or parasitic cause of the diarrhea. Often, these cases are related to a larger food-borne-illness outbreak.

A classic epidemiological investigation begins by verifying the diagnosis through laboratory tests. "If they have not seen a physician, we will ask them to do so, or to submit a stool specimen for examination," Greenblatt says. "We do this fairly early on in the investigative process." The second step is "hypothesis generation"—identifying food sources that may have caused the outbreak. A hypothesis usually grows out of information gathered during the initial interview. The next step is to try to test the hypothesis, usually by having people who were potentially exposed to the same food-borne infection answer a scientifically based questionnaire.

In New Hampshire, health officials recently used this method to investigate a food-borne-illness outbreak affecting ten people who attended the same party. Answers to the questionnaire uncovered many facts, including when the food was delivered, when it was eaten, who ate what, and who the caterer was. Eight partygoers who did not become ill also provided valuable information. As it turned out, all ten members of the sick group, but only three in the nonsick group, had eaten the tossed salad. This provided officials with a statistical association.

Armed with laboratory findings plus the data generated from the questionnaire, a sanitation inspector is deployed to the suspect food-service establishment—in this case, the catering company that provided food for the party. The inspector watches the food-preparation process and takes samples for analysis. The inspector may ask the cook how the raw ingredients were obtained and how the suspect dish was prepared around the time of the outbreak. Greenblatt says that in his experience most food-service establishments go out of their way to assist the investigation. "Occasionally, we do have people who are more adversarial," he says. "They obviously see a threat to their business—and there *is* a threat to their business. I've seen businesses close very quickly because of one case of this or that disease."

Sophisticated laboratory tools are revolutionizing scientists' ability to connect seemingly sporadic cases to a single outbreak. One of these tools is known as polymerase chain reaction, which allows for the recovery of very small quantities of DNA from a stool or food sample. Another is DNA "fingerprinting," technically known as pulsed-field gel electrophoresis. Pulsed-field gel electrophoresis is a simple way of comparing genetic material by cutting the DNA into pieces, then measuring the number and sizes of these pieces. According to Greenblatt, data generated by this technique are usually reliable enough to be admissible in court.

Despite these advances, only about 25 percent of suspected food-borne-illness outbreaks are ultimately confirmed, Greenblatt says. Unlike police who may arrive at a crime scene when the body is still warm, public-health officials seldom find a smoking gun. The interval between the time the contaminated food was prepared and served and the time at which the sickened people or their doctor notified the health department can be a week, sometimes two weeks, the epidemiologist says. By then, the leftovers are in the landfill, and some of the victims have recovered.

While it is not always possible to trace the cause of a food-borne illness, people suffering from gastroenteritis should keep in mind that their malady may be due to a food-borne infection that could quickly progress to a serious illness.

PART 2

KEEPING THE FOOD GROUPS SAFE

4

MEAT AND POULTRY

When *Consumer Reports* announced in February 1998 that two thirds of the one thousand store-bought chickens it tested were contaminated with *Campylobacter,* you may have been surprised, and perhaps a little alarmed—unless, of course, you work in the food-safety arena. To food experts in academia, industry, and government, the study's results were old news. A significant percentage of the raw poultry and meat sold in this country has always been contaminated with disease-causing bacteria. *Campylobacter* infections alone kill at least 500 and sicken an estimated 2 million Americans each year, according to the U.S. Centers for Disease Control and Prevention (CDC). *Campylobacter* can be spread through water and unpasteurized milk, but more than half of campylobacteriosis cases can be traced back to undercooked chicken. Since 1996, *Campy-*

lobacter has established itself as the most common pathogen found in U.S. poultry and the most prevalent cause of food-borne illness in this country.

Reducing the amount of all kinds of pathogens in meat and poultry before these products reach the public is an expensive, complex, and often futile endeavor. The American beef industry reportedly invested at least $8.4 million in food-safety research when *Escherichia coli* O157:H7 was first identified in 1993 as a major threat to food safety. According to the federal government, American taxpayers spend $800 million or more a year on federal food-safety and inspection programs—mostly meat and poultry inspections—that were overhauled and modernized during the 1990s. Central to that overhaul is Hazard Analysis and Critical Control Points (HACCP), the science-based system that, among other things, requires slaughter and processing plants to routinely monitor their products and processes for certain pathogens and to demonstrate the effectiveness of their plans for reducing or eliminating pathogens. HACCP plans vary from facility to facility, but all must be designed to keep contamination within predetermined limits, or baselines, the so-called pathogen-reduction performance standards established by the U.S. Department of Agriculture. Different baselines are set for different species of bacteria and for different classes of foods: cows and bulls; steers and heifers; swine; and chicken and turkey.

HACCP officially took effect in large slaughter plants (those with 500 or more employees) in January 1998, and it will be phased in at medium and small plants by January 2000. Many plants have had HACCP programs in place for several years already. "I think the industry was getting ready for us," says I. Kaye Wachsmuth of USDA's Food Safety and Inspection Service (FSIS). "What will happen over time is that those people who are not meeting the performance standards will either come into compliance or somehow be eliminated."

Just three months after HACCP was officially implemented, the USDA took action that reportedly shut down thirty meat-packing plants or production lines for violating safety standards. According to Agriculture Secretary Dan Glickman, some plants were closed for failing to do required *E. coli* tests, while others failed to follow their own sanitation standards. While circumstances at each plant differed, he said, the fundamental reason they were shut down was the same: the plants were failing to take their new responsibilities seriously enough. Glickman's comments came during a food-policy conference, during which Dane Bernard, food-safety vice president for the National Food Processors Association, said the industry "has taken major steps to reduce contamination," and that more public education efforts were needed to tell consumers how to handle food safely.

The importance of improving safety, in terms of both health and economics, was underscored in August 1997, when Hudson Foods Incorporated was forced to recall 25 million pounds of frozen ground-beef patties and to close its two-year-old, $35 million Columbus, Nebraska, processing plant. Meat from the facility—processed during three days in June of that year—was suspected of causing *E. coli* infection in a number of people in Colorado.

The company, which also processed chicken and pork, came under fire for its practice of using leftovers from one day in the following day's production. Hudson agreed to sell out to Tyson Foods within two weeks of the recall, according to Bloomberg News. It was later disclosed that the deboned beef Hudson had purchased to make its frozen patties may actually have been tainted before ever reaching Hudson's plant. This is not surprising since contamination occurs, in the vast majority of cases, during the slaughtering process. Thus meat can be contaminated at the very beginning of the process, and the bugs just stay with the meat, and may even increase in number, on its way toward becoming hamburger.

Still, the practices at the Hudson plant aggravated the situation. But Hudson was probably doing nothing different from other meat-processing plants. It is just that Hudson got caught. To average consumers, the recall may have generated a feeling of false security. It may have seemed that the government had everything under control—if there is contaminated beef out there, the public will hear about it, and a recall will happen. Unfortunately, nothing could be further from the truth. Later in this chapter you will discover why.

Even if a company follows food-safety rules, the mere perception that its products are contaminated can sound the death knell for an entire corporate entity. A single outbreak, even if it involves only a handful of victims, can trigger multimillion-dollar lawsuits and boycotts.

These enormous financial, regulatory, and public-health pressures to make our meat and poultry safer are beginning to have some impact, some observers say. As Wachsmuth noted, several meat and poultry plants had already brought down their contamination levels before HACCP formally took effect. With the spring 1998 introduction of PREEMPT, a spray-on "cocktail" of competitive bacteria proven to reduce *Salmonella* counts in chickens to almost zero, the task may potentially become easier in the not-too-distant future.

Richard Lobb, spokesman for the National Broiler Council, a Washington, D.C.–based trade group for chicken processors, says data in the *Consumer Reports* study suggest that for chicken, at least, *Campylobacter* and *Salmonella* rates appear to be coming down. Of the 1,000 raw chickens tested by Consumers Union, the publisher of *Consumer Reports,* 63 percent tested positive for *Campylobacter.* Several years ago, Lobb says, *Campylobacter* was detected by other investigators in 90 to 100 percent of raw chickens. He adds that *Salmonella* was found in 35 to 40 percent of chickens during the 1980s, but in just 16 percent of the 1,000 chickens recently tested by Consumers Union. USDA's Food Safety and Inspection Service recently reported that positive tests for *Salmonella* in chicken carcasses fell from 16 percent in 1997 to 9 percent in 1998. Eight percent of the birds tested by Consumers Union were contaminated with both *Salmonella* and *Campylobacter,* and 29 percent tested negative for both. An estimated 32 million chickens are processed every workday in the United States. So even with the recent improvements, at least 3 million contaminated chickens head for public consumption each day.

Despite recent signs of progress, millions of food-borne-illness cases and thousands of deaths each year from complications of food-borne disease are nagging reminders that more work needs to be done. Complicating matters is the emergence of new, more aggressive food-borne pathogens, among them *Salmonella typhimurium* DT104, which some experts fear may be as dangerous or more dangerous than *E. coli* O157:H7. Impervious to such common antibiotics as ampicillin, streptomycin, tetracycline, chloramphenicol, and sulfonamides, *S. typhimurium* DT104 is believed to infect people through consumption of animal products and proximity to animals on the farm.

Some scientists attribute *S. typhimurium* DT104 and certain other strains of drug-resistant bacteria to the widespread use of antibiotics in farm animals. According to the World Health Organization (WHO), the emergence of *S. typhimurium* DT104 is part of a general trend being seen with *Salmonella*. "The incidence of bacterial resistance has increased at an alarming pace in recent years and is expected to continue rising at a similar or even greater rate in the future as . . . antibiotics lose their effectiveness," states a 1997 WHO fact sheet.

Another example of drug-resistant bacteria apparently stems from overuse of fluoroquinolones, a class of powerful antibiotics. Historically reserved only for extremely ill people, two fluoroquinolones won FDA approval for use in poultry in 1995. Many farmers now add these drugs to chicks' drinking water to prevent a disease that could wipe out a flock. According to a recent Associated Press report, Frederick Angulo of the Centers for Disease Control and Prevention says that drug-resistant *Campylobacter* strains weren't detected in humans before 1995, but that 13 percent of human campylobacteriosis cases tested last year were fluoroquinolone-resistant—and the number is rising. According to Minnesota's state epidemiologist Michael Osterholm, constant exposure to fluoroquinolones among chicks on the farm caused *Campylobacter* to rapidly mutate into fluoroquinolone-resistant strains. Of seventy-six chicken products sold in Minneapolis–St. Paul grocery stores in 1997, 79 percent were contaminated with *Campylobacter,* and 20 percent were fluoroquinolone-resistant, the AP reported.

Of the 50 million pounds of antibiotics manufactured each year in the United States, more than 16 million pounds are combined with animal feed or water primarily so that the animals will grow faster, and some 40,000 pounds are sprayed onto fruit trees. Microbiologist Gail Cassell of Eli Lilly & Company notes in the AP story that risk of creating drug-resistant pathogens must be balanced with the realization that antibiotics are vital to animal health. If farm antibiotics are restricted, she says, "the real question is, 'What will it do to the world's food supply?' My plea is that we need more data."

But the FDA considers the threat serious enough that it is preparing stiffer rules for new animal antibiotics, including requiring manufacturers to track treated animals for early signs of resistance, the AP reported.

Zero Tolerance?

It is vitally important to do everything possible to block the emergence of more antibiotic-resistant food-borne bacteria. But is it realistic for the public to expect all the raw meat and poultry they buy to be pathogen-free? Without irradiating the products and opening and preparing them in a sterile environment, complete elimination of food-borne pathogens is an unattainable goal for a variety of reasons:

- The four major bacterial pathogens in meat and poultry—*Campylobacter, Salmonella, Listeria monocytogenes,* and *E. coli*—are ubiquitous in nature, as is the fifth, *Staphylococcus aureus,* a salt-tolerant bacterium that can be a problem in deli-type meats and such products as ham, ham salad, and whole cured sausage. *Staph* can flourish when harmless, competitive bacteria are killed by the curing process.
- Of these five types of bacteria, only *E. coli* O157:H7 is considered an "adulterant"—illegal—and only when detected in ground beef.
- A certain amount of "background" contamination with *E. coli, Salmonella,* and other pathogens is tolerated—and even expected—on solid muscle products.
- At this writing, there is no mandated testing program for *Campylobacter, Listeria monocytogenes,* or *Staphylococcus aureus* in meat or poultry, although some food processors voluntarily

monitor their products for one or more of these bugs.

- There are dozens of areas in slaughter plants and processing plants where microbial contamination can potentially occur.
- Laboratory testing for food-borne pathogens is basically a hit-or-miss endeavor; a carcass or packaged product that tests negative may still contain harmful microorganisms.
- Shiga toxin—a poison produced by *E. coli* O157:H7 that makes people sick—is also produced by fifty to sixty other known bacterial strains. Yet, hardly anyone is testing meat products for the presence of toxin-making bacteria other than *E. coli* O157:H7.
- There is no government oversight of temperatures inside refrigerated trucks hauling perishable foods, such as meat and poultry.
- Food irradiation is rarely used by producers, even though the vast majority of scientific studies have found it to be safe and the most effective means of sanitizing raw meat and poultry before it is sold to the public.

Sara Beck, a technical information specialist at the Food Safety and Inspection Service (FSIS), points out that products such as ground beef, pork chops, ground-turkey patties, and chicken parts that are irradiated or otherwise pasteurized while *inside* the package may become contaminated again as soon as the package is opened. "You'd still have to handle it very, very carefully," Beck says, "otherwise all the efforts to produce a sanitary product will be for naught."

Safety in the Slaughterhouse

In the past the USDA inspected carcasses based on how they looked, smelled, and felt as the carcasses passed by on a rail, conveyor belt, or chain. They looked for observable signs of disease or contamination, such as visible fecal matter. In 1997 the FSIS reinforced its "zero tolerance" standard for visible fecal contamination on livestock and poultry carcasses.

Also, in the past some inspectors paid too much attention to "non-food-safety" issues, such as labeling, which are unrelated to a plant's sanitation or safe food preparation, recalls food technologist Deborah Klein of Greensboro, North Carolina. Klein used to work in a meat-packing plant and later for the USDA, before founding, in 1987, Agribusiness Solutions, a consulting firm to the meat and poultry industries. "Some inspectors would get hung up on the size of a meat cut, or the size of the meat-inspection legend"—that circle with a number inside indicating where the product came from, she says.

While plants are still responsible for accurate labeling, Klein has observed that both federal inspectors and plant managers now pay far more attention to microbiological hazards in food. They look at how an animal is received and held at the plant, and how it is slaughtered, trimmed, washed, cut, put into vacuum packs, prechilled, and loaded onto trucks—with an eye on what the plant is doing to identify and reduce bacterial contamination and growth.

The USDA, which regulates meat, poultry, and shell eggs, can seize ground beef found to be tainted with *E. coli* O157:H7. (The FDA, which regulates almost all other foods, can also seize and destroy contaminated products.) But the agriculture department, at this writing, has no power to order the recall of contaminated meat or poultry; only the company that produced the tainted products can do that. Federal officials can, however, withdraw USDA inspection if contamination at a plant rises to unacceptable levels, and if plant personnel cannot get the situation under control in a reasonable amount of time. In most cases, this would force the plant to suspend operations. But withdrawing USDA inspectors is usually a last-ditch move. "The industry considers our ability to call a quick press conference power enough" to voluntarily hold or recall contaminated products, says Wachsmuth.

The FSIS *Salmonella* testing program, which began in 1996, is designed to ensure that slaughter plants are working to reduce levels of that bacterium. Steps to reduce or remove *Salmonella* from carcasses will also presumably reduce *Campylobacter* and other pathogens since all are associated with fecal contamination. *Salmonella* is being targeted because historically it had been the most common cause of food-borne illness associated with meat and poultry, and because a relatively rapid laboratory test is available.

Under HACCP, slaughterhouses themselves are responsible for monitoring carcasses for generic *E. coli*, another indicator of fecal contamination. (Some also voluntarily conduct their own *Salmonella* tests.)

To conduct the generic *E. coli* test in a slaughter plant, a sanitized sponge is wiped across a small area of the carcass, or a small square of tissue is removed. The specimen is then sent to a certified laboratory. Sampling frequency for the majority of establishments is one test per 300 carcasses for beef, one test per 1,000 carcasses for pork, and one test per 22,000 carcasses for chicken. Low-volume plants slaughtering fewer than 6,000 cattle or 20,000 hogs annually need sample only one carcass per week. Testing such a tiny proportion of a plant's total inventory provides only a rough estimate of actual contamination levels. Indeed, the testing routine is designed more to monitor the processing procedure than to ensure the safety of the final product.

The results of these tests are measured in the number of bacteria, or "colony-forming units" (CFUs) per square centimeter. If none are found, the test is negative; if fewer than 100 CFUs are found, the result is considered marginal; if more than 100 CFUs are detected, the result is unacceptable. If three marginal or unacceptable results are recorded in the last thirteen samples tested, a so-called moving window, the plant must suspend operations so that a HACCP team can review the laboratory data and plant records, and if necessary, comb through the whole operation.

But testing takes time, up to forty-eight hours, although newer tests deliver results in a day. Nonetheless, a large plant could easily process at least 150 carcasses an hour. So by the time a given test result is known, the carcass that was sampled—and the thousands of other animals that shared the same pen prior to slaughter—have already been processed, packed, and trucked out. In some cases, meat suspected of being contaminated has already been consumed by the time the problem is detected and a recall is issued. This happened in the spring of 1998 when IBP Inc. of Joslin, Illinois, the nation's largest meat-packing company, announced a nationwide recall of 282,128 pounds of ground beef. The recall was issued on April 29, more than two weeks after ground beef tainted with *E. coli* O157:H7 had been processed at the plant (on April 14). The recall covered that entire day's production, which had been shipped to twenty-one states. Although IBP Inc. and the USDA were not aware of any illnesses traced to the beef at the time of the recall, the Centers for Disease Control and Prevention alerted health departments in all fifty states to be on the lookout for *E. coli* infections.

The USDA has tested more than 20,000 samples of ground beef since October 1994, and only thirteen have been positive for *E. coli* O157:H7. Whenever a test comes back positive for bacterial contamination, the HACCP team's front-to-back investigation of the plant can usually pinpoint where the contamination is coming from, according to food technologist Deborah Klein. Steps can then be taken to correct the problem and prevent recurrence.

The HACCP team's first order of business is to review pathogen test results and other control data, such as refrigeration temperatures, maintained by the plant. If controls are found inadequate, and if there has been prior serious or repeated HACCP failure, then the plant may be shut down until it can demonstrate to the

USDA that control has been restored and that they can ensure the production of a safe food product.

One of the first control points is the holding pen; the team will want to know whether the animals coming into the holding pens are reasonably clean. Cattle that travel several hours to reach the slaughterhouse will be dirtier than animals trucked in from nearby ranches. Holding pens should also be kept as clean and dry as possible. "The animals are standing out there, crowded into pens with little room to move, so naturally they're defecating on each other or defecating on the ground, and it's splashing up onto their hides," Klein says.

Other points of potential contamination occur when the animal is stunned, killed, bled, and skinned. The investigators may want to know, for example, whether all the employees doing the skinning are sufficiently trained. Are they pulling the hide down and away from the carcass so that the hide isn't touching any of the exposed meat tissue? Do the workers understand why they must skin an animal in such a way so as to avoid contamination? Klein says that employee training is extremely important. "For many workers, English isn't their first language, and they come from different cultures, so their idea of cleanliness may not fit our standards."

Another potential contamination point is the plant floor, which must be cleaned and sanitized periodically as the animals enter the plant and move down the rail. Workers must be extraordinarily careful when eviscerating the animal. It doesn't happen too frequently, Klein says, but when the intestines or stomachs break open and spill their contents onto the carcass, it is considered a bit of a disaster. "The carcass is separated immediately," she says, "and everything's cleaned up and sanitized before the rail can start moving again."

Inspectors also check the cleanliness of the workers' clothing, especially their shoes. For example, employees should not be walking from the holding pens through the processing area. Carcass washes and sprays are investigated; the water must be hot enough and the bactericidal rinses concentrated enough to reduce *E. coli*.

"You really have to become a detective to figure out where the contamination is occurring," Klein says. "You have to look at each step. And there are a lot of players you have to look at, too."

The Broiler Council's Richard Lobb says his industry has also beefed up its sanitation practices over the past several years to combat what he calls perception problems. "For example, plants used to run without chlorination; now they all have chlorinated water, particularly in the chill tanks," he says. Chill tanks are vats of refrigerated water used to reduce carcass temperatures. Additionally, many plants are installing new equipment designed to minimize the amount of bacteria that is transferred from the birds' feathers to their carcasses during slaughter and processing. One device, called an inside-outside bird washer, uses a pressurized nozzle designed to blast contaminants from the surface of the bird and also clean the cavity. More technology is under development to combat *Campylobacter* in chicken-processing plants, he says, adding: "There's a tremendous incentive to get rid of microorganisms, because while you're getting rid of *Salmonella* and

Campylobacter, you're also getting rid of spoilage bacteria."

Clearly, HACCP is heightening the meat-processing industry's awareness of food-borne pathogens and starting to improve sanitation industry-wide. Food-safety experts emphasize, however, that the system is far from perfect. Consumers still need to handle their meat and chicken safely, cook it thoroughly, and make wise choices when patronizing food-service establishments.

Hit or Miss

Even a negative result from the most sophisticated microbiological test method, or "assay," does not guarantee that the food is pathogen-free. Back in June 1997, Hudson Foods reportedly tested for *E. coli* O157:H7 in the batch of frozen patties that was eventually recalled—but missed finding the pathogen. That kind of foible happens all the time, and this is why: Say you buy a pound of ground beef from your local supermarket, take a little one-ounce piece of it, incubate it in broth overnight, look for disease-causing bacteria, and find nothing. This negative finding looks good on paper, but in reality has no bearing on the rest of the gound beef in the package you purchased. The little one-ounce section right next door to the piece you assayed could very well be contaminated, but you wouldn't ever know it. Researchers at Tufts University demonstrated this by examining ten samples from a single pound of ground beef. About half the samples contained disease-causing bacteria; the other half were clean. Microscopic analysis, therefore, provides information about the piece that is actually being tested and offers no guarantees for the remainder of the product.

Some of the Hudson burgers involved in the recall were found to contain only three organisms per 100 grams of meat, and the bugs were not evenly distributed. Because *E. coli* can cause illness at very low doses, the FSIS, in the wake of the recall, increased its ground-beef sampling size from 25 grams to 325 grams per assay.

While testing larger samples may increase the potential of finding *E. coli* O157:H7 in ground beef, there's another problem: laboratories can find only what they are looking for. Unlike fictional "Star Trek" scanners that would give instantaneous readouts of every chemical and microorganism in a food sample, late-twentieth-century technology allows us to test only for specific, known pathogens or toxins—one at a time. So, even if the FSIS were to test every last morsel of a ground-beef patty for *E. coli* O157:H7 and not find it, that doesn't mean you could have eaten the hamburger raw and not gotten sick. *E. coli* O157:H7 is but one of more than fifty known bacterial strains that produce disease-causing Shiga toxin. Another is *Shigella dysenteriae,* from which Shiga's name is derived, although infections with *Shigella dysenteriae* are very rare in the United States. *E. coli* O111:NM also produces Shiga toxin and was implicated in an outbreak in Australia in the mid-1990s. Of 120 Australians who got ill from eating a type of sausage called mettwurst that was contaminated with *E. coli* O111, twenty-two developed hemolytic uremic syndrome (HUS), and one child died. Interestingly, these percentages were similar to those in the 1993 Jack in the Box outbreak of *E. coli* O157:H7 in the Pacific Northwest. Evidentially, *E. coli* O111 has the real potential of causing big outbreaks here. Only no one in the United States is routinely testing for it or for Shiga toxin, even though assays for both exist.

Keeping Cool on the Highway

While the federal government oversees food-safety operations in slaughter and processing plants, it defers to states and localities to regulate food safety in retail establishments. Yet, no governmental entity oversees what happens to food in transit, creating regulatory gaps that you can, well, drive a truck through. These gaps potentially affect almost everything we eat since virtually all foodstuffs in the contiguous United States reach their destinations by truck. "Currently," says Ralph Stafko, a senior policy advisor at the FSIS, "there's no national law that prescribes that products be kept at a certain temperature during shipping."

Take, for example, a semi-trailer hauling raw beef from the Southwest to a restaurant in New York, Philadelphia, or Kennebunkport, Maine. If the truck's cooler does not maintain a steady temperature of 40°F or below, its environment will likely promote the rapid, progressive growth of any pathogenic organisms that might be present on the food. So, a load of raw chicken that is deemed microbiologically safe when it leaves the loading dock at the processing plant can potentially become highly contaminated by the time it arrives at the grocery store. Stafko says he has heard stories about truck drivers turning off their engines to save a few gallons of gasoline and about refrigeration units malfunctioning on the road. "But we don't get hard evidence like who did it, when did it happen, and what happened to that product," he says. "We haven't given up trying to do something about it, but the mechanics of rule making are such that the burden is very much on us to show that there is a definite problem out there before we can make rules that are going to cost people real-life money."

For the most part commercial, contractual, and liability pressures make food safety during transportation a self-regulating enterprise, according to Stafko. For example, contracts between suppliers and shippers, and between shippers and recipients, often stipulate the temperature at which perishable food must be maintained en route. In essence, the contract becomes an extension of HACCP implementation. Some transport companies have built into their trucks devices that measure and document temperature in transit. There are also sophisticated geopositional devices that allow the home company to track not only where their trucks are on the highway, but also the temperature of the food inside. Another approach is to place temperature-indicating devices on selected packages. While these technologies are useful for monitoring and documenting temperature, Stafko says that they are neither required by statute nor universally employed.

At the receiving end, if a supermarket manager or restaurateur rejects a shipment of chicken, pork, lamb, or beef whose safety or freshness is suspect, the hauler and its insurance company lose money. The hauler may try to cut these losses by selling the rejected load to someone else, perhaps at a discount. There are no laws or regulations to prevent this from happening, and the USDA and FDA are concerned. "We don't have a good enough handle on what happens to the rejected product," Stafko says. "Does it go to a third-party salvage operator who sells it to some schlocky outfit who sells to poor people downtown? It's been known to happen."

Another regulatory gap involves "back-hauling"—a practice in which the same truck carries garbage in one shipment and food in the next so that it won't have to hit the road empty. In 1990 Congress passed the Sanitary Food Transportation Act in the wake of several back-hauling scandals. Three years later, the Department of Transportation proposed controversial rules designed to enforce the new law but "basically hasn't done squat since," Stafko says. Claiming it lacked the food-safety expertise to regulate back-hauling, the transportation department unofficially handed the ball to the USDA and the FDA, he says. "The only thing we've actually published to date is an advance notice of rules in November 1996, which covers the basic issues in the Act, truck refrigeration, and our concern about potentially hazardous foods if things are not kept chilled in transit."

Those proposed rules were also met with great controversy. Currently, Stafko says, USDA and FDA officials are conferring with trucking industry representatives with the goal of developing sanitation and refrigeration standards that truckers can live with, perhaps on a voluntary basis.

Food technologist Deborah Klein stresses that truckers, like everyone along the food-supply continuum, play a critical role in keeping the nation's food as safe as possible. "When we're feeding this many people, and food is traveling such huge distances, there is a much greater opportunity for problems to occur, and we have far more challenges than we had back in the old days when restaurants bought meat from local producers," she says. "It's important for consumers to know the whole picture so they can make better judgments—so if the meat on their plate is not served hot, they can think twice about eating it."

Irradiation

In the 1940s, researchers discovered that exposing beef to cobalt-60 gamma rays prolonged the product's shelf life. Since then, irradiation of beef has been endorsed by almost every meat-industry trade group as well as by the World Health Organization, the American Medical Association, and the International Atomic Energy Agency. Some consumer groups allege that irradiation could diminish a food's nutritional value or leave it radioactive, but exhaustive scientific research has shown that when properly administered, irradiation does neither. What irradiation does do is drastically reduce pathenogenic and spoilage bacteria. "This is the most researched technology in food safety," epidemiologist Michael T. Osterholm recently told Bloomberg News. "It's not a panacea, but it's a very important tool."

In December 1997, the Food and Drug Administration gave its final nod to irradiation of fresh and frozen beef, lamb, and pork. The USDA was expected to issue proposed rules on implementing the technology by the summer of 1998. Irradiation has previously been approved for poultry, fruits, vegetables, grains, and spices to control microorganisms, parasites, and insects. Yet only spices are irradiated on a routine basis. Studies have shown that after consumers are educated about irradiation, public acceptance of the technology rises significantly. Some critics disagree, but according to the FDA, irradiation does not noticeably change taste, texture, or appearance of meat and poultry. In some stores where irradiated fruit is displayed beside nonirradiated fruit, most consumers buy the irradiated versions, even though they are slightly more expensive.

But consumers don't have that choice when buying poultry because far less than one percent of the poultry produced in this country is being irradiated. The same story seems to be unfolding for beef. Why irradiation has been so underutilized depends on whom you talk to.

Dean O. Cliver, Ph.D., a food-safety expert at the University of California at Davis, places much of the blame on the federal government for putting out "mixed signals."

"The government grudgingly permits irradiation but has never encouraged it," he says. "They claim it's not their place to encourage any one technology over another. But that strikes me as Pontius Pilate washing his hands. They know good and well that irradiation will do the job, and no alternative is equally certain to do the job. But they're being impartial, just in case."

John Eichberger, manager for public affairs at the American Meat Institute, says that consumer wariness is the main reason his industry has not embraced irradiation. A lack of irradiation facilities is another problem, as is liability. If a meat processor has to send its products to an outside irradiation facility, that processor loses control of its product but is still liable should it become contaminated after being irradiated, Eichberger explains. To solve that dilemma, he says, some companies are looking to develop an "end-of-the-line" irradiation process in which meat, poultry, and pork would be irradiated in the package right before it leaves the processing plant. Years of research and development are needed to make that notion a reality, and Richard Lobb, of the National Broiler Council, balks at the very idea. "An electron-beam emitter," he says, "isn't the kind of thing you can just bolt on to the end of the line. And irradiators use spent rods from nuclear power plants and are best kept in a concrete blockhouse in a big, industrial-size facility. I don't think the people in East Cupcake, North Carolina, would particularly welcome that kind of facility."

At this writing, there is only one irradiation facility in the United States that has been approved for poultry. A company called Food Technology operates it in Mulberry, Florida, and reportedly has yet to make a profit. "It's a shame that so many people have to get sick and die before consumers and companies say yes, this technology should be used," Food Technology President and Chief Executive Pete Ellis told Bloomberg News.

In early 1998, Food Technology's irradiation facility was operating at just 10 percent capacity. Because there is only one facility, and it is so far away from the bulk of America's two hundred chicken-processing plants, using it would be a logistical nightmare, according to Lobb.

Meanwhile, a group of fast-food restaurant chains is jointly studying the feasibility of using irradiated beef and chicken in their menu items, says Kim Miller, corporate spokesperson for Burger King. A subcommittee of the National Council of Chain Restaurants is exploring a number of irradiation issues, including whether large enough irradiation facilities could be built to handle the tremendous amount of food sold by fast-food outlets. "If we were to adopt irradiation, we'd want to make sure that our entire supply went through that process," Miller says.

A second area of study is what irradiation would do to the cost of fast-food. A third issue is how irradiation might change the meat's appearance, texture, and chemical compounds. "Then there's that other looming question: How will consumers react to a product that has been irradiated?" Miller says. "This process may prove to be very viable, but at this point, there are still an awful lot of questions remaining."

No Home for Mad Cows?

While irradiation may someday solve the problem of new and emerging pathogens in our meat and poultry supply, there is one food-borne illness that the USDA says meat-eaters in America probably need not fear: a form of Creutzfeldt-Jakob disease (CJD). CJD is a rare, always-fatal neurological disease and the human equivalent of bovine spongiform encephalopathy (BSE), or mad cow disease.

First described in the United Kingdom in 1986, BSE has been traced to a newly recognized infectious agent called a prion found in cattle remains that are fed to cattle, which by nature are herbivores. Through November 1996, about 165,000 head of cattle in Britain were diagnosed with mad cow disease. BSE manifests as a chronic neurological ailment that causes paralysis; infected cows stagger and fall over. In humans, CJD is a degenerative brain syndrome marked by dementia and sudden muscle contractions. Most victims die within a year after symptoms begin, but not until after losing their coordination, vision, speech, intellect, and personality.

During a 1996 segment on mad cow disease, talk-show host Oprah Winfrey declared she would never eat another hamburger. Soon afterward, cattle sales allegedly fell to ten-year lows. A Texas cattlemen's organization subsequently sued Winfrey. In February 1998 a jury vindicated Winfrey even though, thus far, there has never been a confirmed case of BSE in the United States. In an effort to keep it that way, the USDA in 1989 banned importation of all cud-chewing cattle (ruminants), bovine semen, embryos, meat, and bone meal from the United Kingdom. That prohibition was extended in 1997 to prevent imports of live ruminants and most ruminant products from Europe. A total U.S. ban on cattle protein in cattle feeds for domestic animals became fully effective in October 1997. By then, many cattlemen had already halted the practice voluntarily.

Considering CJD's years-long incubation period, it remains to be seen whether these actions will protect American citizens from contracting the disease from BSE-infected beef. Dan McChesney, Ph.D., leader of the animal-feed safety team at the FDA's Center for Veterinary Medicine, is optimistic. "Our approach has been very proactive; it was put in place with no history or indication that BSE is in this country," McChesney says. "The risk has always been very small, and we're taking measures to make it even less of a risk."

In the spring of 1998, the USDA entered into a cooperative agreement with Harvard University's School of Public Health to begin an analysis and evaluation of the department's current measures for preventing BSE. According to a USDA press release, the two-year study will review current scientific information, assess the ways that BSE could potentially enter the United States, and identify any additional measures that could be taken to protect human and animal health.

Meat and Poultry Safety Tips

The USDA's Food Safety and Inspection Service offers the following tips for safely handling meat and poultry. (More guidelines, including cooking temperatures and storage charts, appear in Chapters 9 and 10.)

- Do not defrost foods on the kitchen counter or anywhere other than in the refrigerator, in cold water, or in the microwave oven.
- Meat and poultry should not be brought to room temperature before cooking. Bacteria can grow when food temperature is in the "danger zone" of between 40° and 140°F.
- Wash hands before and after handling raw meat or poultry.
- Thoroughly wash the food processor, meat grinder, cutting board, utensils, and other surfaces after they come in contact with raw meat or poultry.
- Marinating time in the refrigerator should not exceed recommended storage for fresh meat or poultry (three to five days for red meat and one to two days for poultry). Marinades impart flavor; they do not destroy bacteria. Never marinate at room temperature.
- Use a meat thermometer to ensure that meat is thoroughly cooked.
- Use a rack when cooking whole poultry so hot air can circulate under it.
- Do not truss whole poultry legs; fold wings akimbo to facilitate heat getting into joints for more even, thorough cooking.
- Use a minimum oven temperature of 325°F for cooking meat and poultry, or casseroles containing them.
- Do not partially cook or brown foods to cook later because any bacteria present would not have been destroyed.

- Do not use low-temperature, overnight cooking except in a slow cooker.
- Meat or poultry going into a slow cooker or Crock-Pot should be cut into chunks or small pieces to ensure thorough cooking.
- Turn or rotate food during the cooking process, especially when using a microwave oven.

Daniel Y. C. Fung, a professor of food science at Kansas State University, says the following test can be used to determine whether a piece of meat or poultry has been fully cooked. Cut off a small piece of meat and drop it into a spoonful of 3 percent hydrogen peroxide. If you see tiny bubbles, it means the meat has not been heated to a temperature high enough to destroy any bacteria that may be present.

5

SAFETY FROM THE SEA

Carol's company party was an elegant affair. It took place in a fancy New York City dining club and featured a buffet of exotic cuisine from around the world. The most popular table held an elaborate raw bar, complete with a glistening mermaid ice sculpture and Russian caviar. Carol's husband and coworkers loaded their plates with raw oysters, clams, and mussels. Carol took a half-dozen cooked shrimp and a dollop of cocktail sauce from bowls beside the mermaid's tail.

"Come on," her husband cajoled, as he motioned to put some shiny bivalves on Carol's plate. "How many times a year do we get to eat raw oysters? It's a delicacy."

"I don't think so," replied Carol, who had recently read about a man with suspected hepatitis who'd died ten days after eating raw oysters. She knew that her husband, as well as others at the party, had also read that story, which had appeared on the front page of *The New York Times* several days earlier. Reading about the potential danger of *Vibrio vulnificus*—a disease-causing bacterium that may be present in raw shellfish—did not seem to influence anyone else's eating behavior at the party. But Carol had been spooked by that story. She vowed never to eat shellfish again unless it was thoroughly steamed, baked, or fried.

As it turned out, neither her husband nor any of her coworkers became sick as a result of their indulgence. Chances are that Carol also could have enjoyed the raw bar with impunity. If she had gotten sick, her healthy immune system would probably have kept her from becoming seriously ill.

If Carol was being overcautious, she is in good company. "I don't know any food microbiologist who will eat raw shellfish," says food microbiologist Tom Montville of Rutgers University. "But a lot of people eat it, and nothing ever happens. There is plenty of safe raw shellfish, but you never know."

Since 1989 at least 167 *Vibrio vulnificus* illnesses traced to raw oyster consumption have been reported in the United States; of those cases, 87 people died. On one hand, that's a good safety record considering that the FDA estimates more than 20 million Americans eat raw oysters. On the other hand, when *Vibrio vulnificus* invades the bloodstream, it is fatal about 50 percent of the time. *Vibrio vulnificus* infections produce such symptoms as vomiting, diarrhea, abdominal pain, fever, and chills. Fatalities usually occur when the patient goes into septic shock. In almost every fatal case, the victim had underlying liver disease or a weakened immune system. In one vibriosis outbreak affecting 16 people in Los Angeles who had eaten raw oysters, 12 had preexisting liver disease associated with alcohol use or viral hepatitis. Having abnormally low stomach acid and taking antacids also raise your risk for severe vibriosis symptoms.

Oysters, mussels, clams, scallops, and cockles are "filter feeders," which means they obtain sustenance by filtering out algae and other food from the water. It is estimated that two to three gallons of water pass through a mollusk's body every hour. Unfortunately, these shellfish also indiscriminately filter out and concentrate in their bodies any industrial pollutants, viruses, bacteria, or other unwelcome microscopic entities that happen to be floating by. When you eat the animal inside the shell, you are consuming the entire organism, including its version of a gastrointestinal tract.

While *Vibrio* infections from raw oysters account for a significant number of seafood-borne diseases reported each year in this country, seafood can carry a variety of other hazards that either occur naturally or are introduced by humans. Raw shellfish can harbor Norwalk viruses or hepatitis A if the animals were mishandled after harvest or grown in waters polluted with human effluent. Other microorganisms from sewage pollution include *Salmonella typhi* (which causes typhoid fever); and two strains of *Vibrio cholerae* (which cause gastroenteritis). Microorganisms such as *Salmonella, Shigella,* and *Staphylococcus aureus* may be spread to seafood by handlers, equipment, or the environment.

Toxic levels of histamine can accumulate on tuna,

mahimahi (dolphin), bluefish, and other members of the scombroid family that are not kept cold enough after capture. Ciguatoxin, which causes gastrointestinal and neurological problems, is found in some tropical fish species, including amberjacks, barracuda, snappers, and groupers. Another naturally occurring toxin, domoic acid, may be present in bivalve shellfish (oysters, clams, and mussels) that filter out one-celled plants called diatoms, which produce the acid. Parasites, such as tiny roundworms and tapeworms, may be naturally present in the flesh of freshwater fish. Another source of shellfish contamination are naturally occurring algae blooms called red tides. The FDA and the coastal states test for these blooms, and when they appear, the waters are closed to all shellfish harvesting.

In general, thorough cooking will destroy bacterial and viral pathogens; freezing fish at –4°F for seven days or longer will kill parasites; but neither freezing, cleaning, nor cooking will make toxic fish safe to eat.

The most perishable of all flesh foods, seafood needs to be treated with the utmost care from the time it lands on a fishing boat until it reaches the dinner plate. For the most part, the domestic seafood industry and the estimated 135 foreign countries that export seafood to the United States have a good safety record. However, every illness and every death that results from eating contaminated seafood is evidence of a weak link in the safety chain—a link that can usually be strengthened through education or better policing of the industry.

HACCP

Perhaps the most common public misconception about the seafood industry is that it is not policed or regulated at all. This, says seafood-safety specialist Donn R. Ward, Ph.D., of North Carolina State University, is simply not the case.

"Certainly one could argue that seafoods do not receive the same level of carcass-by-carcass inspection that consumers may have knowledge of in the meat and poultry industry, where a USDA inspector looks at every chicken, hog, beef carcass that comes by and makes some sort of determination," Ward says. Nor is there an FDA inspector on site every day at all 5,000 to 6,000 seafood-processing plants in the United States. A federal inspector may visit a plant only once a year, or once very five years at low-risk operations, although state inspectors visit much more frequently.

The FDA has, however, taken a number of steps geared toward preventing contaminated seafood from reaching the public. For example, the FDA administers the National Shellfish Sanitation Program, which helps the twenty-three shellfish-producing states and nine shellfish-exporting countries control sanitation during growing, harvesting, shucking, packing, and interstate transportation of clams, oysters, and mussels. The FDA also helps train state and local health officials who inspect fishing areas, seafood-processing plants, restaurants, and other retail outlets, notes a recent article in *FDA Consumer* magazine.

On December 18, 1997, the FDA tightened oversight of the seafood industry when it began requiring all seafood processors to implement a Hazard Analysis and Critical Control Points (HACCP) plan. HACCP applies to domestic seafood processors as well as exporters to the U.S. market. The FDA estimates that about half of America's seafood supply is imported from such places as Canada, Mexico, Ecuador, Russia, and China. According to the *FDA Consumer* article, food companies in the European Union, Canada, and some other foreign countries already use HACCP, and the FDA is working on international agreements with other countries to follow suit.

While the Environmental Protection Agency and other agencies routinely test fishing waters for pollutants and natural contaminants, the federal government recommends, but does not require, microbial testing of seafood under HACCP, as it does for chicken and meat. The sheer volume of seafood and the speed at which it perishes make routine testing unworkable, says Robert Collette, director of food regulatory affairs at the National Fisheries Institute, a trade group based in Arlington, Virginia. "There really is no way that you could do enough testing to be absolutely certain that a pathogen isn't there, anyway."

HACCP is nonetheless having a major impact on how seafood is being harvested and handled after harvesting, observes Laura Garrido, seafood nutrition specialist with the Bureau of Seafood and Aquaculture at the University of Florida. "The word is getting out," she says. "There are more people with more seafood-safety education working in the business. This HACCP has opened their eyes. Seafood used to be safe, but now it's even safer."

Seafood-borne Hazards Introduced by Humans

Familiar terrestrial food-borne pathogens, such as *Salmonella* or *Campylobacter,* are not part of the natural intestinal flora of seafood. When fish become hazardous to human health, it is usually because the water they lived in was contaminated or because the fish were mishandled after capture. Seafood HACCP plans are therefore focusing mainly on water quality, holding temperatures, and sanitation practices. Among other things, HAACP is requiring seafood producers to keep very detailed records of where and when their fish were captured. Processors are also responsible for temperature control. "HACCP is a fundamentally different approach for an industry that has been regulated, but not highly regulated, in the past," says Ward, who has trained seafood producers and regulators in HACCP principles. "It's not business as usual."

While HACCP does not cover commercial fishing vessels—there are far too many boats to make enforcing HACCP practical—officials are hopeful that the new regulations will ultimately improve onboard safety controls. "The FDA is basically banking on the processor extending responsibility of product safety back to the harvesters," says Ward.

Critical control points—the points at which hazards can be prevented, eliminated, or reduced to acceptable levels—vary according to which species of fish is being processed and what the end product is going to be. For molluscan shellfish, water quality is a major control point, particularly if the mollusks are destined for a raw bar. Processors must document that their shellfish came from approved waters.

Fishing may be restricted in certain areas for a variety of reasons, not just the high *Vibrio* levels that are common in the Gulf of Mexico and other warm waters during the summer months. Problems may also occur in waters because they are close to sewage plants, industrial development, marinas, or large farming operations where terrestrial bacteria associated with human and animal waste can get into the water column. While the vast majority of shellfish harvesters and processors follow the law, Ward says there are bootleggers who will unscrupulously harvest from prohibited areas at night and may try to sell the goods from the side of the road. "It is unwise for consumers to buy from people whom they don't know to be reputable dealers," he cautions.

Problems can also occur in waters that are not restricted. In 1993 there was an outbreak of Norwalk virus gastroenteritis associated with eating raw oysters gathered from a remote oyster bed. According to a report by CDC investigators published in the *Journal of the American Medical Association,* crews from twenty-two of twenty-six oyster-harvesting boats working in the same area routinely disposed of sewage by dumping it overboard. That sewage probably included contaminated feces from one or more ill harvesters. The investigators concluded that education of oyster harvesters and enforcement of regulations governing waste disposal by oyster-harvesting boats might prevent similar outbreaks.

Recreational Fishing

HACCP does not extend to this country's very substantial recreational fishery, which is undoubtedly contributing to the number of seafood-borne illnesses reported each year. While some commercial fishermen are seeking out training in HACCP principles on a volunteer basis, this is not happening to any measurable extent among recreational fishermen. "Recreational fishermen might be doctors, lawyers, or Maytag repairmen with no clue about seafood safety," Ward says. "If we were shooting cattle and chickens from the interstate and putting them in our trunks, you'd see a lot more food-borne disease outbreaks associated with meat and poultry."

When recreational fishermen catch more tuna or mackerel than they can use, they may share it with friends and neighbors, or try to sell it to restaurants. Or, if they catch a tuna that is bigger than their ice chest, they may stick it in the shade on their boat for half a day instead of throwing it back. By sundown, there may be enough histamine built up in the tuna flesh to sicken everyone who consumes it.

Coolers on fishing boats should have at least one pound of ice for each pound of fish stored in it. Three inches of ice should cover the bottom of the cooler and another three inches should lie on top of the fish; ice should also be interspersed between each fish. After unloading, the cooler should be thoroughly cleaned, rinsed with chlorinated water, and restocked with fresh ice to prevent bacterial buildup.

Seafood suppliers know better than to try to sell

illegally caught or improperly stored fish to Legal Sea Foods Inc., a Boston-based restaurant chain, according to Stephen Martinello, Legal Sea Food's director of quality control. Martinello conducts laboratory tests on specimens from every seafood shipment the company receives. He performs fecal coliform analysis on raw molluscan shellfish that indicates whether the mollusks were grown in the presence of sewage effluent. He also tests various seafood products and salads for Shiga toxin, which is produced by some *E. coli* strains, including the notorious *E. coli* O157:H7. Every load of tuna, mahimahi, bluefish, and others in the scombroid family is checked for histamine levels. *Staphylococcus* analyses are done on precooked crabmeat and crab claws. Crabmeat is also tested for the presence of *Listeria monocytogenes.* Cream soups are tested for pathogens, as well.

If a tainted product is found, Martinello immediately sends the shipment back to the distributor and reports his findings to his friend, the shellfish officer for the Commonwealth of Massachusetts. Martinello estimates that Legal Sea Foods spends at least $200,000 a year on its laboratory, testing equipment and supplies, and quality-control personnel, who in addition to testing products, also inspect sanitation procedures in each of the chain's fourteen restaurants on a monthly basis.

"It's a large investment, but it translates into customer confidence," says Martinello, a registered sanitarian and registered environmental-health specialist. "Whenever there's any type of bad publicity about seafood or shellfish, our stores seem to be busier than other stores." In the twelve years Martinello has worked for the company, there have been no illness outbreaks traced to Legal Sea Food restaurants, nor has he heard of any outbreaks at Legal before he joined the company.

University of Florida researchers are developing an "electronic nose" that can sniff out odor molecules on seafood to help companies like Legal Sea Food decide whether to accept or reject a catch. But Martinello says he doesn't expect he'll need the $40,000 device, even though it reportedly is almost 100 percent accurate. "If you're in the fish business and can't pick up a piece of fish and look at gills, nape, eyes, and smell it, feel it, and know whether it's fresh," he says, "you should be selling hamburgers."

Other Seafood-borne Diseases

Scombroid Fish Poisoning. The most critical control point in avoiding scombroid fish poisoning is temperature. Mackerel and related saltwater fish species should be kept cold continuously throughout the harvesting, processing, warehousing, and retailing phases. If certain species, such as tuna, mahimahi, mackerel, bluefish, sardines, anchovies, and bonito, are not iced soon after death, their flesh will begin to break down. This breakdown creates increasingly high levels of histamine, a chemical that can cause severe allergic reactions in humans. Even if temperature was properly controlled until a fish reached a retail outlet, histamine can amplify to dangerously high levels if the consumer or eating establishment fails to refrigerate the fish promptly after purchase. Histamine levels can build up before the fish appears spoiled or smells bad.

Scombroid fish poisoning can be explained by looking at the saltwater fish's body chemistry. Typically, Ward explains, if you put something in salt water, the salt will migrate into the product and make it salty. But a saltwater fish does not taste salty because when it is alive, its body manufactures small chemical compounds called osmoregulators. Osmoregulators float around the fish's tissues and counter the pressure of salt, keeping it out. In scombroid-susceptible species, one of the osmoregulators is an amino acid called histidine. After the fish dies, spoilage bacteria convert histidine to histamine. "Eighty degrees on the deck of a boat for four or five hours is ample time for this phenomenon to occur," says Ward. Histamine is heat stable; it is not destroyed by cooking.

Ward notes that in most cases, histamine levels in illness-causing fish have been above 200 parts per million (ppm), and often over 500 ppm. Large tuna canners routinely check for histamine, but this has not been true for histamine-susceptible species marketed in other forms (i.e., fillets). An elevated histamine level alone does not alter a fish's smell or appearance. Under HACCP, processors do not have to check for histamine as long as they can document that the product has been chilled very soon after harvest and maintained under refrigerated conditions.

If the processor cannot demonstrate onboard temperature control through appropriate documentation from the harvest vessel, then the FDA may require that the processor analyze a sample of fish to show that histamine levels are below 50 ppm. LeeAnn Applewhite, manager of seafood diagnostics for Neogen Corporation of Lexington, Kentucky, says she gets numerous calls from processing plants soon after they are inspected by the FDA. "If their records aren't 100 percent verifiable all the time, the FDA says they have to set up a plan to test for histamine," Applewhite says. She acknowledges that the FDA's recent crackdown has been good for her company, which sells a variety of food-safety test kits. Processors must also document that no more than 2.5 percent of their fish have off odors or show any other signs of decomposition.

The presence of osmoregulators such as histidine is one reason seafood is more perishable than other types of flesh foods. It also explains why catfish, trout, shad, and other freshwater species, which do not have osmoregulators, have a slightly longer shelf life than saltwater fish.

Ciguatera. Because marine creatures exist in a fish-eat-fish world, it can lead to a rather unique type of food poisoning called ciguatera. The initial culprits are a species of microscopic plankton, or dinoflagellates, which produce ciguatoxin. Like histamine, ciguatoxin is heat stable and cannot be detected by taste or smell. These natural toxins can become concentrated as they move up the food chain, but their adverse effects appear to be limited to humans. Almost any tropical fish around Florida and the Caribbean has the potential of becoming ciguatoxic, but the most common are amberjacks and other jacks, moray eels and barracuda, hogfish, scorpion fish, certain triggerfish, and certain snapper and grouper species.

Call a doctor or the public-health department if

you develop gastrointestinal symptoms followed by such neurological problems as headaches; flushing; muscular aches; dizziness; or tingling and numbness in the lips, tongue, and mouth less than six hours after eating tropical fish. Try to get portions of the meal—particularly the fish—pack it tightly in ice, and freeze it for later analysis. If you can get information on how the fish was handled before you ate it, it will aid in determining whether ciguatoxin is actually to blame. Signs of poisoning usually subside within several days.

Parasites. Parasitic worms can be contracted by eating raw or lightly cured freshwater fish that was not previously frozen to kill parasites. Fear of parasites is one reason many people avoid sushi and sashimi, Japanese dishes that feature raw fish on vinegared rice. But according to a fact sheet from the University of Delaware Sea Grant program, fewer than thirty cases of illness resulting from parasites in sushi or sashimi were reported in the United States in 1996. Most of these cases occurred on the West Coast because Pacific mammals such as seals, porpoises, and whales are the most common carriers of parasites.

Trained sushi chefs know how to detect parasites through a procedure called candling, or examining the fish through a light. Infected fish should either be discarded or frozen so as to kill the parasites. The freezing temperature needed to kill the parasites is lower than that obtainable in the average home freezer. Freezing will not, however, kill bacteria. For that reason, people with diabetes or other medical conditions that place them at high risk for food-borne diseases should avoid eating raw fin fish as well as raw shellfish.

Avoiding Cross-Contamination

At her local grocery store's seafood counter, Debbie orders a pound of raw sole fillets and a pound of cooked shrimp. She is pleased to see that the man behind the seafood counter is wearing plastic gloves as he weighs out the fillets. But her pleasure turns to horror as he grabs the cooked shrimp out of the case without washing or changing his gloves first. When Debbie points out his folly, the man tells her to wash the shrimp at home if she is so concerned. But Debbie isn't satisfied. She takes her complaint to the store manager, who orders the employee to throw out both items and start over, this time washing his hands and changing gloves between weighing out the raw and cooked products.

The cross-contamination mishap Debbie witnessed represents one of several ways ready-to-eat seafood products can become vehicles for food-borne illness. According to the Florida Cooperative Extension Service, other ways to cross-contaminate seafood include:

- Storing precooked seafood in the same container previously used for raw seafood or other uncooked foods;
- Handling ready-to-eat seafood with utensils, gloves, or work surfaces that were previously used for raw seafood; and
- Temporarily storing or thawing ready-to-eat

items in a sink or container previously used to hold raw seafood.

Despite efforts among harvesters, producers, and government regulators to make seafood products safe, retailers and consumers must still presume that all raw seafood is contaminated and handle it as carefully as they would handle raw meat or poultry. Retailers, for instance, should not display raw and ready-to-eat seafood side by side. And raw seafood should not be placed above cooked seafood on an ice mound; melting ice can carry bacteria from the raw to the cooked product.

Washing fish steaks and fillets that you buy from a grocery store or fish market is not necessary; the practice may do little more than contaminate your sink. Washing cross-contaminated cooked seafood will not entirely eliminate pathogens, either, according to seafood-safety specialist Donn Ward. Recooking the product would do the trick, but double cooking might also eliminate the edibility characteristics for which you bought the product in the first place.

People who enjoy the taste of raw oysters but want to minimize their health risk can ask retailers or restaurants about a new pasteurization process called Ameripure. By exposing the oysters to warm and ice-cold waters in a systematic way, the Ameripure Oyster Company of Empire, Louisiana, says it is able to reduce *Vibrio vulnificus* and most other bacteria to nondetectable levels. The process adds six to eight cents to the wholesale cost of an oyster. According to a recent story in *The New York Times,* some people find that Ameripure oysters taste just like regular raw oysters;

others find them less juicy and a bit rubbery. Ameripure oysters are currently sold in restaurants in twenty-six states. To find out which restaurants serve them, consumers can call the company toll free at 800-EAT-M-RAW (328-6729).

The Bottom Line

In Ward's view, efforts to reduce seafood contamination under HACCP will not lead to any "dramatic decreases" in seafood-borne illness outbreaks. He does, however, expect to see some decreases over time. A more obvious and more immediate side benefit to HACCP may be an overall improvement in seafood quality. The steps needed to improve safety also serve to keep seafood fresher. "One of the biggest hurdles the seafood industry has had to face over the years is problems with inconsistent quality," he says. "The closer to fresh a product is, the better your eating experience is going to be in terms of flavor, texture, aroma, and appearance."

Seafood Safety Tips

The following consumer tips offered by the FDA are designed to ensure that your seafood is safe and as fresh as possible:

SHOPPING

- Fresh seafood should not smell unpleasantly "fishy." It should smell like a "fresh ocean breeze."
- Fresh fish steaks and fillets should be moist, with no drying or browning around the edges. The eyes of fresh whole fish should be bright and clear, not cloudy or sunken. Scales should not be slimy and should cling tightly to the skin. Gills should be bright pink or red. Frozen fish should not be freezer burned or have damaged packaging.
- Mollusks in the shell should always be alive when you buy them. When a clam, oyster, mussel, or scallop is alive, the shells will be tightly closed or will close when tapped lightly or iced. A test for freshness is to hold the shell between your thumb and forefinger and press as though sliding the two parts of the shell across one another. If the shells move, the shellfish is not fresh. Throw away any that do not close tightly.
- Buy seafood only from reputable dealers. You cannot know what you are buying from the back of a pickup truck. It could have been caught by someone not subject to FDA or state inspection.

STORING

- Keep fresh fish in the coldest part of your refrigerator, usually near the freezer or in the meat drawer, until it is ready to cook and serve.

- Store fresh fish in your refrigerator in the same wrapper it had in the store.
- Store live mollusks in your refrigerator in containers covered loosely with a clean, damp cloth. Do not store live shellfish in airtight containers or in water.
- Canned fish should be refrigerated after opening.
- Smoked fish, pickled fish, vacuum-packed fish, and modified-atmosphere-packed fish products should always be refrigerated.
- The safest way to thaw frozen seafood is in the refrigerator in its own container. Allow about one day for defrosting.

COOKING

- Cook fish no later than two days after purchase.
- For fin fish (baked, broiled, poached, fried, or stewed): allow ten minutes cooking time for each inch of thickness. Turn the fish over halfway through the cooking time unless it is less than a half-inch thick. Add five minutes to the total cooking time if the fish is wrapped in foil or cooked in a sauce. Properly cooked fish will flake easily with a fork and should be opaque and firm, not translucent.
- Boil live oysters for three to five minutes after the shells open.
- Steam shellfish four to nine minutes from the start of steaming.
- Use small pots to boil or steam shellfish. If too many shells are cooking in the same pot, it is

possible that the ones in the middle won't get thoroughly cooked. Discard any clams, mussels, or oysters that do not open during cooking. Closed shells may mean they have not received adequate heat. For shucked oysters, boil or simmer for at least three minutes or until edges curl. Broil three inches from heat for three minutes. Fry in oil for at least ten minutes at 375°F. Or bake (as in oysters Rockefeller) for at least ten minutes at 450°F.

6

Raw Fruits, Vegetables, and Juices

Now that you know how difficult it can be to remove pathogens from raw meat, poultry, and seafood, you may be entertaining the idea of becoming a vegetarian. But before you start downloading meatless recipes from the Internet, take a moment to consider a raspberry. Imagine that this sweet, aromatic fruit was grown in a pristine, American field protected from roaming deer and livestock and fertilized with properly composted manure. Imagine that it was drip irrigated with purified water, minimally treated with pesticides and herbicides, cultivated by a healthy farmworker who had just put on a pair of sanitized rubber gloves, gently placed in a disinfected bin, washed in chlorinated water, and shipped to its destination in a clean truck with perfectly calibrated temperature and moisture controls.

Imagine this berry arriving at a store that happens to follow the best HACCP plan ever devised. And imagine that you buy, rinse, and eat the raspberry at its peak of ripeness.

All this, and the berry can still make you sick.

People can control many aspects of growing fruits and vegetables and bringing them to market, but who controls the wildlife? And who controls other people's personal hygiene habits? No one, for example, can completely eliminate flies that can potentially transport pathogenic bacteria from fresh animal dung to an unprotected raspberry patch. No one can stop a blue jay or swallow from swooping over a field of raspberries and dropping pathogen-laden waste from twenty feet above. No one is watching the field-worker or grocery-store stock clerk in the bathroom to make sure he or she washes hands after using the toilet. No one may stop a child from running his or her filthy hands along a swath of raspberries in the grocery store. And only Mother Nature controls the raspberry's intricate architecture, which provides a perfect hiding ground for microscopic organisms. To be sure, the only raspberries associated with illness in recent outbreaks in the United States were grown in Guatemala, and produce contamination from bird feces is only theoretical. However, should a raspberry pick up contamination in the field; on a truck; in a store, restaurant, or someone's kitchen, the only way for mere mortals to make it safe without cooking it is to scrub it until it is destroyed.

There are many hoops that a raspberry or any raw fruit or vegetable must jump through before reaching the human palate, and each hoop presents a potential for contamination.

Doctors and nutritionists stress that the health benefits of eating plenty of fruits and vegetables each day far outweigh the small risk of contracting a produce-borne infection. And most food-safety experts agree that the vast majority of produce sold in the United States is safe, particularly after it has been thoroughly washed.

However, over the past decade, the rising popularity of a produce-rich diet has led to an increasing number of food-borne illnesses attributed to contaminated produce and fresh fruit juices. For example, outbreaks of salmonellosis associated with fresh fruits and vegetables traditionally have been rare, notes a 1994 report written by Minnesota health officials and published in the journal *Clinical Infectious Diseases*. Yet, from 1990 through 1993, four multistate outbreaks of salmonellosis, each involving 100 to 400 confirmed cases, were attributed to fresh fruit or vegetable sources, the authors point out.

In recent years, outbreaks have been traced to *Shigella,* hepatitis A, *Cyclospora,* and *E. coli* O157:H7 on lettuce; *Campylobacter* and *Salmonella* on cantaloupes; *Salmonella* on tomatoes; *Salmonella* and *E. coli* on alfalfa sprouts; *Cyclospora* on raspberries; *E. coli* and *Campylobacter* on melons; hepatitis A on frozen strawberries; and *E. coli* O157:H7 and *Cryptosporidium* in fresh, unpasteurized fruit juice. Poor food-handling practices are usually, but not always, to blame.

The major source of produce contamination comes from direct or indirect contact with infected human or animal wastes, according to the Food and Drug Administration (FDA). In a 1997 report on produce handling and processing practices, Larry R. Beuchat and Jee-Hoon Ryu of the University of Georgia at Griffin note that *Listeria monocytogenes, Clostridium botulinum,* and *Bacillus cereus* are naturally present in some soil and that their presence on fresh produce is "not rare."

"Soil on the surface of fruits and vegetables may harbor pathogenic microorganisms that remain viable through subsequent handling to the point of consumption unless effective sanitizing procedures are administered," Beuchat and Ryu write in a recent issue of *Emerging Infectious Diseases,* which is published by the Centers for Disease Control and Prevention (CDC). Contact with mammals, reptiles, fowl, and insects offers another avenue through which pathogens can reach produce, according to the report. The report further states that *Salmonella, E. coli* O157:H7, *Campylobacter jejuni, Vibrio cholerae,* parasites, and viruses can contaminate fresh produce through contact with raw or improperly composted manure, irrigation water containing untreated sewage, or contaminated wash water. In 1990 and 1993, two salmonellosis outbreaks involving a total of about 300 cases in four states were traced to consumption of fresh tomatoes. Tomatoes from both outbreaks were traced back to a single packing house where a wash tank appeared to be the likely source of contamination.

In 1995 an outbreak of *E. coli* O157:H7 infections involving at least twenty-nine people was linked to lettuce. While it is not known where the lettuce became contaminated, the people who investigated the outbreak noted that the lettuce had been irrigated with surface water, which may be vulnerable to contamination possibly through runoff, according to the FDA. *E. coli* O157:H7 was again identified as the culprit in a 1996 outbreak in which unpasteurized apple juice sickened seventy people in three western states and killed a little girl in Colorado. The juice maker, Odwalla of California, allegedly obtained the tainted batch of juice from apples that may have been contaminated with deer feces, although this was never proven. The company now pasteurizes its juices.

In another widely reported outbreak that same year, as many as 1,465 people in twenty states, the District of Columbia, and two Canadian provinces were sickened from Guatemala-grown raspberries that were contaminated with the parasite *Cyclospora cayetanensis*. The growers' water supplies were suspected, but precisely how the berries became contaminated has not been determined. Two similar diarrhea outbreaks traced to Guatamalan raspberries occurred in 1997 and 1998. In 1997 frozen strawberries tainted with hepatitis A, which attacks the liver, caused almost 200 Michigan schoolchildren to become sick. Health officials traced the outbreak to fruit grown in Mexico, but they still don't know where or when the hepatitis A virus attached itself to the strawberries.

Although some of the most highly publicized outbreaks have been traced to imports, epidemiologist Michael T. Osterholm of the Minnesota Department of Health says there are no reliable data to show that imported produce presents a bigger risk than domestically grown produce. "Any food item that is not going to be cooked is potentially a highly vulnerable food, whether it's grown in the U.S. or not," he maintains.

Although less than 2 percent of confirmed food-borne-illness outbreaks have been attributed to raw fruits and vegetables, some researchers have suggested that fresh produce may be the single biggest cause of food-borne disease. Consumers' failure to wash all produce before eating is a contributing factor in many of these outbreaks.

In 1996 an illness outbreak in New York, Connecticut, and Illinois landed at least twenty-one people in the hospital and almost killed three-year-old Haylee Bernstein of Connecticut. The outbreak was traced to California-grown, precut, prewashed lettuce leaves contaminated with *E. coli* O157:H7. Because the leaves were labeled "prewashed," Haylee's mother reportedly didn't bother to rinse them.

Felicia Busch, M.P.H., R.D., a spokeswoman for the American Dietetic Association, says, "You should wash absolutely everything from bananas to zucchini." Wash bananas? "Think about how a child eats a banana," she says. "They put their hands all over it to peel it; then they use those same hands to eat it. So the possibility of cross-contamination with either bacteria or pesticide residues is just tremendous with young kids." Giving bananas a scrub before peeling, and washing hands before eating, can reduce the risk of infection.

Even if produce is safe when you buy it at the store, you can easily create a problem by getting raw meat or poultry juice on salad greens or by touching fresh produce with unwashed hands. "I saw a kid sneeze on a grape table in the grocery store a couple of weeks ago, and ever since then, I've washed every piece of produce I buy just because you don't know who's handled it," says Julia Daly, a spokesperson for the U.S. Apple Association.

Certain fruits and vegetables—including lettuce, raspberries, and strawberries—seem to be more bacteria-prone than other produce items. Take lettuce. On the farm, lettuce picks up bacteria from the soil. While most of these bacteria are innocuous spoilage bacteria, some may be pathogenic. Even if the lettuce is labeled prewashed and looks clean in the grocery store, bacteria still remain—10,000 or more per gram, by some estimates. Bulk lettuce can also be contaminated in the grocery store; it is sold out in the open, so anyone can touch it or sneeze on it. You can lower your infection risk by rinsing bulk lettuce under running water two or three times before serving.

While there have been no reported illnesses (at this writing) traced to prewashed "ready-to-eat" bagged lettuce, that does not mean it has never happened or never will happen. Bagged lettuce sold in grocery stores is primarily consumed in home settings—venues that lend themselves to sporadic, hard-to-detect outbreaks. As with most perishable foods, lettuce is likely to spoil before any pathogens that happen to be there can multiply to dangerous levels. However, in the context of diseases like *E. coli* O157:H7—where as few as fifty bacteria can be a lethal dose—why take any chances? Bagged lettuce should be rinsed at least once, especially if it is to be eaten by someone at high risk for food-borne illness.

Sprouts, meanwhile, fall into a category all their

own. For most foods, processing makes them safer. But sprouts are unique in that their processing actually enriches any pathogens that may be present instead of killing or reducing them, says Linda J. Harris, Ph.D., microbial food safety specialist at the University of California at Davis. In order to sprout, the seeds must be incubated under moist, room-temperature conditions for three to five days, Harris explains. This, she says, is "a perfect environment for the multiplication of pathogens even if present on the seeds at undetectable levels."

Michael Osterholm says he won't eat uncooked sprouts because in his opinion there are no sufficiently effective means of ensuring the sanitation of the seeds or the delicate tendrils that grow out of them. Treating the seeds with irradiation or chemical disinfection inhibits the sprouting process. Sprouted seeds seem to defy disinfecting, as well. Japanese researchers recently applied a strong disinfectant to *E. coli*–contaminated radish sprouts for ten minutes, but even that failed to eliminate the bacteria growing below the sprouts' surface. In experimentally contaminated radish sprouts, at least, the researchers reported that *E. coli* "exists deep in plant tissues where the disinfectant could not reach." In 1997 *E. coli*–tainted alfalfa sprouts sickened at least seventy people in Michigan and Virginia. The CDC said the sprouts were probably contaminated by animal waste while they were still seeds.

"I can't just tell consumers, 'If you cook your [salad] to 160 degrees, you can solve any potential problem,'" says Caroline Smith DeWaal, director of food safety for the Center for Science in the Public Interest. "We laugh, but in Japan when that *E. coli* O157:H7 outbreak from sprouts was going on a couple of summers ago, they cooked their salads. They cooked everything because they didn't know the source of the bacteria." The seeds reportedly had been exported from the United States and incubated in Japan.

Sprouts have become staples in salad bars, which Osterholm calls "one of the best laboratories in the world" for microbial contamination and transmission. You can have the very best system for growing and distributing produce—right to the point of delivery to a five-star restaurant. Then a kitchen worker sticks his dirty, ungloved hand into a bag of prewashed lettuce. One thoughtless act negates everything that was done earlier to make the lettuce safe.

In the late 1980s, Suzanne B. and her mother, Jane, may have been victims of such thoughtlessness when they became seriously ill after eating in a New Jersey restaurant. The only food they ate in common was the dinner salad, recalls Suzanne. "I remember my mother commenting that the lettuce was warm, but we didn't think anything of it," Suzanne says, acknowledging that she has no proof that the salad was contaminated. Within twenty-four hours, both were violently ill with unstoppable diarrhea. Suzanne's stomach was in such a state of distress that she ate nothing but tea and toast for the next seven days. The episode occurred ten years ago, she says, "but I remember it like it was yesterday."

The old meat-and-potatoes diet of a generation ago may have caused more heart attacks, Osterholm says, but it didn't cause as many food-borne diseases. Just how many food-borne infections are occurring as a result of contaminated produce is impossible to know because the vast majority of cases are never reported, and most that are reported cannot be confirmed. But many experts agree that the number of cases has been rising in recent years because of increased produce consumption and the globalization of the produce market. According to the USDA's Economic Research Service, the amount of fresh produce consumed by the average American has gone up about 50 percent since 1988, to about 300 pounds a year.

Imported Produce

Imports have risen dramatically in recent years to satiate Americans' year-round appetite for fresh fruits and vegetables. That demand is driven by more and more Americans wisely following heart-healthy, cancer-fighting diets, which are rich in fruits and vegetables.

Shipments of all imported foods have doubled since 1991, and imports now account for 38 percent of the fruit and 12 percent of the vegetables consumed in the United States. In the winter months, as much as 70 percent of the produce sold in certain regions of the United States is grown offshore, mostly south of the border. In addition to outbreaks from Guatemalan raspberries and Mexican strawberries, other serious outbreaks linked to imports include cholera from frozen coconut milk processed in Thailand, and *Cyclospora* infection from lettuce grown in Peru.

In early 1998 the CBS news magazine *Public Eye* broadcast an exposé of potentially hazardous farming practices in some parts of Mexico. Chemical pesticides that are banned in the United States are used liberally and legally in some developing countries, the *Public Eye* report asserted. Another concern is that some off-season strawberries and tomatoes sold in America were harvested in Mexico by workers who defecate in the fields because there are no toilet facilities provided for them.

In October 1997, a few months before that report aired, President Clinton announced an initiative to upgrade domestic food-safety standards and to ensure that fruits and vegetables coming from overseas are "as safe as those produced in the United States." Specifically, foreign growers seeking to sell produce in the U.S. market will have to meet certain criteria for sanitation in the field and in production plants, as well as with regard to worker health and water quality. Under his proposal, the FDA would be able to halt imports of food from any foreign country with food-safety systems that do not provide the same level of protection required for U.S. products. In March 1998, Clinton followed up that announcement by proposing that the Food and Drug Administration budget be increased by $25 million in Fiscal Year 1999 to hire about 250 additional inspectors.

According to Representative Anna Eshoo, D-California, these changes are necessary. FDA inspections of food imports have dropped 50 percent since 1992, says Eshoo, a sponsor of a House bill aimed at improving safety of imported produce. Today the FDA inspects just 1 or 2 out of every 100 imported food shipments arriving at U.S. ports, and only a tiny fraction of samples undergoes laboratory analysis for disease-causing microorganisms.

"There is a massive effort under way within the

FDA, the USDA (U.S. Department of Agriculture), the EPA (Environmental Protection Agency), and the CDC to work together to really improve the safety of fruits and vegetables," says Arthur Whitmore of the FDA's Center for Food Safety and Applied Nutrition.

Daly, of the U.S. Apple Association, says the feeling in her industry is that the President's initiative is politically motivated. "Produce is extremely safe," she says, "but it's sort of like the last food frontier that hasn't been scrutinized yet." Daly admits, however, that there have been enough outbreaks associated with raw produce and raw juice that regulators and producers need to pay more attention to safety issues.

Unlike animal products, raw fruits and vegetables will not be subjected to periodic laboratory testing for pathogens because it is impractical, Whitmore says. "For juice, it's easier to work with. But if you have five tons of apples, where do you start?"

Sarah Delea, vice president of communications for the United Fresh Fruit and Vegetable Association in Alexandria, Virginia, says that microbial testing of fresh produce is also unreliable. "Testing, in some cases, can be like searching for a needle in a haystack," Delea says. "It doesn't really assure safety."

She points to the Guatemalan farmers who opened their fields to inspectors from the FDA and CDC after their *Cyclospora*-tainted raspberries caused disease outbreaks in the United States. But despite a full investigation, the source of the elusive parasite could not be pinned down. "How did *Cyclospora* get on raspberries when it didn't get on other products that were coming out of Guatemala that were harvested around the same time?" Delea asks. "This is a prime example of why we need more funding for research to develop a better understanding of parasites like *Cyclospora.*"

There are, however, some simple control strategies that can reduce the risk of contamination of imported produce. For example, shipping fruit from Central America with clean ice or in closed refrigerator trucks, rather than with ice made from untreated river water, is "common sense," Robert V. Tauxe, M.D., of the CDC writes in a recent issue of *Emerging Infectious Diseases.*

HACCP

Hazard Analysis and Critical Control Points (HACCP) covers produce canners and has been proposed for makers of raw juice, but as of this writing, there was no HACCP mandate in the pipeline covering the fresh fruit and vegetable industries. "HACCP is based on science," says Delea. "Because the data out there for fresh fruits and vegetables is so limited, you can't really put together a HACCP plan because you don't have the data to support it."

As an alternative, federal regulators have developed a set of voluntary guidelines called Good Agricultural Practices (GAPs) and Good Manufacturing Practices (GMPs). These preventive strategies are aimed at domestic and international growers, harvesters, handlers, and transporters of fresh fruits and vegetables. Part of the President's Food Safety Initiative, the GAP/GMP program focuses on a variety of areas, including improving methods for detecting *Cyclospora* and hepatitis A on produce. The program also addresses water quality, manure management, and sanitation in the field and production facilities, as well as worker hygiene. Research on technologies and strategies to reduce or eliminate microbial contamination is a specific priority. Final GAP/GMP guidelines were expected to be published in the *Federal Register* in late 1998.

A thirty-four-page draft of those guidelines issued by the FDA in April 1998 identified human and animal feces as the main sources of contamination as produce is grown, picked, or processed. The draft guidelines urge growers to monitor their workers' health and to make sure they are trained in basic hygiene, such as washing their hands with soap and water after using the toilet and covering any lesions or wounds that might come in contact with produce.

The draft guidelines were criticized by some consumer groups because they did not carry the force of law. And Stacey Zawal, of the United Fresh Fruit and Vegetable Association, asserted that most food-borne-disease outbreaks happen further down the distribution line because people preparing food are not properly washing their hands. "That is not necessarily true for growers and packers," she told reporters after the draft guidelines were made public.

The FDA guidelines notwithstanding, Delea, of the United Fresh Fruit and Vegetable Association, says the fresh produce industry has been actively and aggressively working to ensure the safety of its products. In the summer of 1997, for example, twenty produce organizations developed a "guidance document" identifying four major risk areas: water; manure; worker, field, and facility sanitation; and transport and handling. Delea says the document puts forth a "common-sense approach" to reduce microbial contamination.

Daly, of the U.S. Apple Association, says all raw apples and other tree fruit—known collectively as fresh market items—are picked off the tree, brushed, and washed before being shipped out for sale. Once they are picked, they aren't touched again by human hands until they reach the grocery store. Pickers are admonished against harvesting "drops"—fruit that has dropped off trees—and putting them into fresh-market bins.

Pesticides

In the 1960s and early 1970s, Americans were quite concerned about pesticides, such as DDT, being used on food crops—and justifably so. Many of the agricultural chemicals used then were fat soluble, which means they could not be removed by water; they accumulated in both the plants and the people who ate the plants.

Over the last two decades, the whole agrochemical industry has been moving away from that type of product, says Barry Swanson, Ph.D., a professor of food science at Washington State University. Modern pesticides are, for the most part, water soluble and thus relatively easy to remove through washing. Pesticides are also used in "very trace quantities," he says. "There have been a few problems, but it's not extensive." He points out that most of the concern expressed in the scientific literature cites foreign countries that legally use pesticides that are banned in the United States.

In the United States, Swanson says, pesticides on produce are not considered a food-safety problem, "at least not by knowledgeable people." Chemicals that fall under the broad heading of pesticides are used to control insects, rodents, weed, mold, mildew, and fungi. Most are not, however, formulated to destroy disease-causing bacteria, protozoa or toxins.

Irradiation

Delea says irradiation is one technology producers are investigating as a means of destroying biological hazards that may be clinging to products as they leave the production plants. Irradiation has been used successfully on strawberries on a limited basis but, according to Delea, has not been widely adopted by the industry for several reasons. Certain produce items, such as melons, cucumbers, and lettuce, cannot tolerate irradiation; Delea says the process makes them mushy. And historically, the industry was wary concerning consumer acceptance of irradiated foods, although this is beginning to change. "While people like the idea of irradiation to pasteurize produce, they don't like the idea that it extends shelf life because they want everything fresh," Delea says. "A lot of people like the idea of going to the market once or twice a week to get fresh produce." Consumers who want irradiation can talk to their grocers about it, she advises, adding that if enough people ask for irradiated fruits and vegetables, producers will provide it.

Felicia Busch, of the American Dietetic Association, says she'd like to see irradiated produce made available as an option, especially for groups at high risk for food-borne illnesses. Ironically, these same high-risk groups—young children, pregnant women, people with AIDS or another immune deficiency, and cancer patients—are urged to eat lots of fresh fruits and vegetables. "Those are the groups that need them the most."

Another sanitation tool is "ozonation," or pumping large quantities of ozone gas into water that is used to rinse produce in the production plant. Producers are also trying a variety of disinfecting washes.

Organic Produce

Often for health and environmental reasons, some consumers turn to organically grown produce. But the same potential for pathogenic contamination exists here, too. Noting that organic farms rely heavily on manure as fertilizer, some food-safety experts express concern that this practice could potentially contaminate crops because fresh manure is a proven reservoir for harmful microorganisms. But Bill Wolf, president of the Organic Materials Review Institute, based in Eugene, Oregon, says that certified organic farms—those that sell at least $5,000 worth of produce annually—must follow specific guidelines for manure use. For example, they must either compost the manure until it has reached a temperature of 150°F before adding it to soil, or they must work the manure deeply into the soil a minimum of two months prior to harvest. Wolf says that either of these methods kills pathogens or gives beneficial bacteria enough time to edge out disease-causing bacteria.

According to Wolf, this country's 12,000 organic farmers—who constitute a $3.5-billion-a-year business—are urging retailers to buy organic produce only from certified organic farms that follow the aforementioned manure guidelines. If you patronize small organic farms that sell through roadside stands or farmers' markets, Wolf suggests talking to the grower about his or her fertilizing methods.

If you have your own vegetable garden, don't fertilize it with fresh manure. Read the label on store-bought manure to make sure it has been adequately composted to destroy pathogens. And don't forget to thoroughly wash or scrub, when possible, all your organic produce.

Ways to Wash

Swanson says washing produce in warm water works better than cold to loosen and remove dirt or soil. While dirt can be microbiologically clean, it may also indicate the presence of invisible microbial contamination, and extra care should be taken to remove all soil from produce, especially if it is going to be consumed raw or lightly cooked. Produce washed with warm water should be cooked or eaten right away.

Washing also removes most or all of the waxy coating that some producers put on their fruits or vegetables to extend shelf life. (Federal law requires retailers to label waxed produce.) Some of the waxes contain antimicrobial agents, usually a fungicide to prevent mold and rot, but the coatings could possibly trap pathogens on the fruit's surface, notes Mindy Brashears, Ph.D., a University of Nebraska Cooperative Extension food-safety specialist. "That's an area that hasn't been thoroughly explored scientifically."

Berries can be cleaned by letting them soak in a bowl of cold water for five minutes, then rinsing them under running water. The rinsing step is very important, stresses extension specialist Linda Harris. "When you place produce in a tub of water, some bacteria will fall off the surface, contaminating the surrounding water," she explains. "This essentially spreads the microorganisms around more evenly, contaminating all of the produce in the bowl since water clings to the produce as it is removed." Harris says that rinsing washes off the bacteria that are loosely clinging to the fruit. Soaking in a dilute bleach solution—one teaspoon of bleach per gallon of water—will kill those bacteria that fall off in the water.

Susan Sumner, Ph.D., a food microbiologist at Virginia Polytechnic and State University, has devised a disinfecting system that anyone can use at home. The system uses two agents: 3 percent hydrogen peroxide solution (the standard strength that is sold in stores to disinfect cuts and scrapes) and vinegar (mild acetic acid). Put each into its own spray bottle. Sumner's research has found that misting with hydrogen peroxide and vinegar in either order can reduce the level of pathogens from the surface of most produce items. The system is not perfect, but it can help make fresh produce safer, she says. After misting, you may rinse the produce under running water, although Sumner says sensory studies found the sprays do not significantly affect taste, especially if you will be using a vinaigrette dressing on your salad. The pair of sprays also can be used to sanitize counters, cutting boards, and other food-preparation areas, she says. The hydrogen peroxide should be changed daily because it breaks down and loses disinfecting power soon after leaving its original container.

Of course, some people view Sumner's system as overkill.

"Usually anything that's left on produce is in such tiny amounts that if, once in a while, you don't follow the rules, it's probably not going to make a big difference over the long haul," says Busch, of the American Dietetic Association. "What I worry about are families that let the kids eat grapes and things like that unwashed at the grocery store, and then don't wash anything when they bring it home."

Produce-Safety Guidelines

Producers play a very important role in reducing pathogens on fresh fruits and vegetables during growing and harvesting. But many infections linked to fruits and vegetables have resulted from cross-contamination in the kitchen coupled with temperature abuse (failing to keep cold foods cold and hot foods hot). Consumers can minimize their risk of eating contaminated produce by taking a few simple precautions:

- Refrigerate fresh produce in loosely sealed plastic containers or plastic bags to prevent cross-contamination in your refrigerator. The American Society for Microbiology says the number of pathogenic bacteria decreases on vegetables when stored at temperatures just below freezing.
- Buy tofu only from refrigerated display cases and be sure the product is cold to the touch. Refrigerate promptly at home.
- Wash your hands before preparing, serving, or eating fresh produce.
- Do not keep cut or cooked produce items at room temperature for more than two hours.
- Do not eat bruised produce.
- Be sure your cutting board, knife, vegetable brush, and peeler are sanitized before coming in contact with produce. You can clean utensils in a mild bleach solution (1 tablespoon per half-gallon of water), in the dishwasher, by drenching them with boiling water, or by hand with dish detergent and hot water.
- Do not peel or cut into a produce item that has not been washed. If the outside of the produce item is contaminated, your knife or peeler may transfer pathogens from the surface to the inside of the fruit or vegetable.
- Before cutting, slicing, or peeling, wash all fruits and vegetables for at least one minute under running tap water. This includes whole melons, cantaloupes, and kiwis. Any produce item that can withstand peeling—including carrots, celery,

and apples—should also be able to withstand scrubbing with a vegetable brush.

- Do not forget to scrub the blossom end of tree fruits—the end opposite the stem—because this is where bacteria like to hide.
- Unless they are to be eaten right away, cantaloupes and honeydews should be refrigerated before and especially after they have been sliced.
- Wash precut lettuce, whether it is sold in bulk (spring or mesclun mix) or in a sealed plastic bag.

Raw Juice

The practice of pressing raw apple juice, or apple cider, in the United States began hundreds of years ago when the first European settlers brought apple presses with them on their ships. Historically cider was thought to be a safe agricultural commodity because it was considered acidic enough to kill harmful bacteria and even some spoilage bacteria. But as the Odwalla outbreak so tragically demonstrated, *E. coli* O157:H7 is one bug that has, as Daly puts it, "jumped the fence."

Raw juice is vulnerable to outbreaks for four main reasons: 1) pathogens on a few pieces of fruit can spread throughout an entire batch of juice; 2) low-level contamination can cause severe illness, especially in young children and other high-risk populations; 3) certain pathogens can survive acidic conditions in juice at refrigerator temperatures; and 4) raw juice is not subjected to a kill step, such as pasteurization, which destroys or greatly reduces contamination. According to the FDA, more than 98 percent of all fruit and vegetable juices sold in the United States are pasteurized.

The CDC has identified two possible causes of raw apple juice contamination: use of "drops"—apples that have been blown off trees into fecally contaminated soil—and contaminated water, usually from wells or lakes.

The Odwalla outbreak prompted most, if not all, apple-producing states to strengthen oversight of cider mills in their respective jurisdictions. Some states have beefed up safety regulations for inspecting, washing, and brushing apples before they are pressed. States are also urging or requiring a ban on using drops to make cider. Any fruit that touches the orchard floor can potentially become contaminated, particularly from the fecal matter of such animals as deer, which can be difficult to control. Many states are also requiring mills to analyze water potability several times a year. These are among the issues to be addressed through HACCP programs, which the FDA was expected to propose for the raw juice industry in the latter part of 1998.

Purifying cider with ultraviolet light instead of heat could be in widespread use by the 1999 cider season, Daly predicts. It turns out that *E. coli* O157:H7 dies when exposed to UV radiation. A $6,000 portable UV unit developed at the New York State Agricultural Experiment Station at Cornell University pumps a thin film of cider past a UV light at a rate of about three gallons per minute. Tests indicate that the machine reduces *E. coli* O157:H7 contamination from 100,000 microorganisms per milliliter (ml) of cider to 1 organism per ml in a single pass. Food safety expert Randy Worobo, who worked with two engineers to design the unit, says the machine should be ideal for small cider producers. Further tests are needed to determine whether UV radiation is equally effective in eradicating other types of pathogens. However, even with these types of safeguards, it is important to keep juice refrigerated so as to prevent the minimal number of remaining bacteria from multiplying back up to harmful levels.

In the meantime, the FDA, along with state health departments, and the U.S. Apple Association have been working to educate cider producers nationwide about sanitation and contamination issues, pasteurization, and HACCP. Daly says that the education process was an uphill battle in some cases because cider mills are relatively small operations run by people who tend to be entrenched in their ways. "We're saying, 'Look, you're running into something your grandfather never even dreamed about. You need to respond accordingly and change your practices,' " Daly relates.

She says many producers are responding very responsibly and want to know how they can protect themselves, their families, and their customers, especially the children. "But," she says, "I am disappointed to say that there is a part of the industry out there that still doesn't get it and is not taking the type of steps we would like to see to effect some change here." Issues of contention are primarily economic; not using apples that have dropped to the ground reduces income generated by the orchard; pasteurization units cost upward of

$30,000—beyond the reach of some small cider presses. There is also an assumption that all pasteurization methods destroy cider's full-bodied flavor, which is not necessarily the case, according to Daly.

Also generating friction among producers is a new FDA requirement to place a warning label on unpasteurized cider. The label must state that the product may contain pathogens known to cause serious or life-threatening illnesses, that the juice has not been processed to destroy such pathogens, and that the risk of serious illness is greatest for children, the elderly, and people with weakened immune systems. "That's pretty rigid," Professor Swanson of Washington State University said shortly before the labeling rule was approved. "The industry doesn't like the idea of having a label on its product saying it may be injurious to health." A compromise might be to post a warning notice at checkout counters where raw juices are sold, he suggests.

According to Safe Tables Our Priority (STOP), warning labels are important. The group says Odwalla was one of at least five identified outbreaks since 1991 that were traced to raw cider contaminated with *E. coli* or *Cryptosporidium*. Citrus juice manufacturers are not immune to problems, either. According to *The Great Lakes Fruit Growers News*, an on-line trade publication, contaminated orange juice has produced outbreaks of gastroenteritis, typhoid fever, and hepatitis A. A salmonellosis outbreak in 1995 proved to be the first such occurrence from a citrus-processing facility, the *News* reported in its September 1997 issue. The bacteria came from tiny tree frogs that were found around the orchard and in the processing facility.

Juice Safety Guidelines

Microbiologist Susan Sumner says she drinks unpasteurized cider herself but would be hesitant to give it to her three-year-old daughter. If you are not in a high-risk group for food-borne illness, you can reduce your small but real risk of becoming sick from contaminated juice by following these simple steps:

- Heed the "use by" date on the bottle.
- Avoid buying cider that is not refrigerated and cold.
- Keep raw cider refrigerated at home.
- Do not leave juice out at room temperature after it has been poured—especially if it is not pasteurized.

To further reduce your illness risk:

- You can pasteurize raw cider at home by heating it to boiling, or to at least to 180°F, for one second. Heating raw cider may alter the flavor a bit, but you can make the beverage tastier by adding some allspice or a cinnamon stick, suggests Daryl Minch, an educator with Rutgers Cooperative Extension. Heated cider should be drunk hot or refrigerated promptly to avoid recontamination.

- Be cautious about patronizing raw juice bars because their juices are not subjected to a kill step. STOP advises children, seniors, the immune-impaired, and pregnant women to refrain from drinking raw juice, no matter where it comes from.

7

EGGS AND DAIRY PRODUCTS

For Denise and her husband, Ben, the new refrigerator was an impulse buy. Theirs was working fine, but the gleaming white, top-of-the line appliance that caught their eye was bigger than the tired, brown model in their kitchen. The new refrigerator sported glass shelves instead of coated wire ones. It dispensed water and crushed ice from the freezer door. Its automatic ice maker would settle the couple's weekly spat over who forgot to refill the ice-cube trays. Denise especially liked the nifty egg holder—a clear, plastic container with a cover and lift-out tray that could hold more than a dozen eggs. The biggest selling point for Ben was the design of the refrigerator door: Its bottom shelf was big enough to hold two one-gallon jugs of milk plus a quart of coffee creamer.

Denise and Ben might regret their new purchase when they discover that some of the conveniences built in to the new refrigerator fly in the face of food-safety concerns. Take the oversized milk shelf on the door. David K. Bandler, professor of food science at Cornell University in Ithaca, New York, says the door is the worst place in the refrigerator to store milk, which is highly perishable. For one thing, the door is farthest away from the cold air source. Also every time you open the door, you expose that milk to the 70-degree air in your kitchen, and you sweep that air back into the refrigerator. "The milk is 'outside' the refrigerator longer than you think," Bandler says. This happens because the normal 30-degree temperature difference between the inside and outside of the refrigerator dips to a difference of about 10 degrees each time you open and close the door. The more frequently you open and close the door, the longer it takes for the refrigerator to cool back down. Likewise, milk takes much longer to cool down than it does to warm up. When a product warms up, bacteria can grow, notes Bandler, who specializes in dairy-processing science.

After conducting experiments aimed at establishing the speed at which various foods spoil, Bandler and his colleagues devised this simple rule of thumb: Every 5-degree increase in temperature cuts a product's shelf life in half. So, if you store milk at 40 degrees, it will stay fresh for ten days, but it will last only five days if you store it at 45 degrees. On the other hand, if you keep milk at 35 degrees, it will last twenty days.

Milk will stay coldest if you store it on a shelf in the refrigerator box—as far back as possible. The door should be used for ketchup, mustard, soda, and other less perishable items.

Eggs should also be stored on a shelf, and in their original carton—not in a built-in egg bin or some fancy plastic holder with a lift-out tray. Picking up eggs and putting them in another container has the potential of transferring bacteria and viruses from your hands to the eggshells, explains Betsy Crosby, a home economist for poultry programs at the U.S. Department of Agriculture's Agricultural Marketing Service. Also, moving eggs around can create minute cracks through which surface bacteria can enter. "The egg carton is designed to protect that little egg and keep it clean during transportation from the processing plant to the store, and from the store to your home," Crosby says. "So why put it at risk?"

While the Food and Drug Administration, at this writing, does not mandate HACCP for the dairy-processing industry, the agency does set very strict limits on the number of microorganisms that can be present in dairy products. These limits are part of the Pasteurized Milk Ordinance, which Bandler says is designed to protect all grade-A dairy products. The FDA enters into contracts with state agencies, such as New York's Department of Agriculture and Markets, to administer the ordinance, Bandler says, adding that the dairy industry "operates under more rules than HACCP ever thought about. . . . The dairy industry has been very closely monitored for a long time."

All that regulation does not let consumers off the hook, however. A basic understanding of how eggs and dairy products can become contaminated before and after they reach your kitchen can motivate you to take steps to prevent many food-borne diseases.

Shell Eggs

There are several possible ways in which pathogens can infect eggs. Disease-causing bacteria from the hen's fecal matter can be transferred from the feathers to the shell after the egg is laid. This usually isn't a problem because fresh shell eggs are washed and sanitized before they are sold. Even if some pathogens linger, the egg whites or yolk probably won't become contaminated unless pieces of contaminated shell fall into the egg after it is cracked open, and the egg is then consumed raw or undercooked. Bacteria on the egg's exterior may also be sucked down into an uncracked egg through tiny pores in the shell. Scientists are trying to figure out exactly how this happens, Crosby says, but it seems to occur only if the pathogen load on the shell is very high and the egg is kept too long at room temperature, or there is too much moisture on the egg. And, like any other food item, cooked eggs can become cross-contaminated if they come in contact with a contaminated plate or utensil.

Of greatest concern, however, is intrinsic, internal contamination of shell eggs with the bacterium *Salmonella enteritidis* (SE). Historically, the inside of an egg was considered practically sterile. In recent years, however, SE infection primarily from undercooked eggs and poultry has become one of the leading causes of food-borne diseases in this country. A recent USDA report on egg-safety strategies estimates that 2.3 million of 46.8 billion shell eggs produced each year in the United States are infected with SE, resulting in an estimated 661,663 illnesses annually. These result in about 3,300 hospitalizations and 390 deaths. SE from all food sources sickens 1.8 million to 2.5 million Americans each year and costs up to $3.6 billion in medical expenses, lost wages, and loss of life, according to previous estimates from the federal government.

Internal SE contamination of eggs begins when *Salmonella enteritidis* from the gut of a laying hen invades the bird's bloodstream and lodges in its ovaries and oviducts, the tubes that carry eggs out of the hen's body. The SE bacteria can then infect the developing egg before the shell is formed.

First identified in the Northeast, SE-tainted eggs have cropped up in other parts of the country, as well. Between 1976 and 1994, the proportion of reported *Salmonella* infections that turned out to be SE increased from 5 to 26 percent, according to the CDC's National Salmonella Surveillance System. From 1985 to 1995, U.S. states and territories reported 582 SE outbreaks, which accounted for 24,058 cases of illness, 2,290 hospitalizations, and 70 deaths—mostly of elderly nursing-home residents. At least four reported SE outbreaks during 1994 and 1995 were traced to the consumption of raw shell eggs. Like those who develop other types of food-borne illnesses, the vast majority of people who contract SE do not seek medical attention.

For reasons that are not entirely clear the threat is highest in the Northeast, where it has been estimated that 1 in 10,000 eggs is internally contaminated with

SE, according to the CDC. According to the American Egg Board, a commodity-promotion organization whose activities are overseen by the USDA, the overall SE contamination rate is estimated to be 1 in 20,000 eggs.

According to the federal government, the average person in the United States eats 200 shell eggs a year. Depending on what part of the country you are in, the likelihood of your finding an infected egg may be as low as one in a million, according to the Egg Board. In the Northeast, your chances may be 100 times higher; about 1 in 50 consumers will encounter a contaminated egg each year in that part of the country. A contaminated egg is made safe by thorough cooking.

Salmonella enteritidis outbreaks are usually traced back, not to individual eggs prepared in home kitchens, but to pooled eggs—large numbers of eggs that are removed from their shells and blended together. Egg pooling may be done at a processing plant, or in a restaurant or institutional setting that feeds omelets and other egg dishes to large numbers of people. Just as one *E. coli*–tainted apple can contaminate an entire batch of cider, one bad egg can potentially contaminate a batch of pooled eggs, especially if the egg pool's holding temperature is not kept at 40°F or below. The potential danger of pooled eggs was driven home during a nationwide outbreak of *Salmonella enteritidis* in the fall of 1994. The outbreak was traced to some Schwan's ice cream that had been made from a base or "premix" that was transported in the same tanker-trailer that had carried pooled raw eggs immediately before. An estimated 224,000 people across the United States were stricken with gastrointestinal illness after eating the contaminated ice cream. The scope of that outbreak is not surprising if you believe a 1997 study in the *Journal of Food Protection* that found that ingestion of no more than 28 SE cells was enough to make someone sick. (Other studies have found the infectious dose of SE to be 10,000 to 100,000 cells.) According to the CDC, each time 500 eggs are pooled in a bowl, there is a one-in-twenty chance that one of them is contaminated.

Clearly, the safest way to eat shell eggs is to prepare them individually, cook them thoroughly, and eat them promptly. But many people prefer their eggs soft boiled or sunny-side up, and some like to eat raw eggs, even though the nutritional content of raw and cooked eggs is the same. (In a scene from the 1976 movie, *Rocky*, actor Sylvester Stallone drinks a glassful of raw eggs before working out.) Arlene L., an artist from New Mexico, remembers blending a raw egg or two into her children's milkshakes to make the treat more nutritious. "I would never do that today," says Arlene, now a grandmother.

According to Edward S. Josephson, Ph.D., an adjunct professor in the Department of Food Science and Nutrition at the University of Rhode Island, the problem of SE in eggs could be eradicated by irradiation. In February 1998, Josephson petitioned the Food and Drug Administration to allow irradiation of fresh shell eggs to rid eggs of any *Salmonella* that might be present. Josephson's research team has demonstrated that gamma rays from cobalt-60 safely destroy pathogenic bacteria without altering the flavor or texture of an egg. He says gamma rays from cesium-137, X rays, or electrons from a linear accelerator—the common irradiation methods for food processing—would probably work equally well. If people could buy irradiated eggs, Josephson says, "they would be able to eat a soft-boiled egg, poached egg, egg in a cup, or even raw eggs, and kids would be able to lick cake or cookie batter off the spoon, without the anxiety of possibly contracting a food-borne disease." In such homemade foods as Caesar salad dressing, hollandaise sauce, mayonnaise, ice cream, and eggnog, irradiated raw eggs could enhance the culinary experience more safely.

At this writing, the FDA was still reviewing Josephson's request. In a letter to the agency, Josephson pointed out that every year of delay contributes to the 9,000 deaths and 30 million or more illnesses from food-borne infections. He says he hopes his letter will confer a sense of urgency. But even if egg irradiation became legal tomorrow, implementation wouldn't happen overnight, if ever. The same logistical, financial, and public-relations dilemmas that have stopped the poultry and produce industries from going full steam ahead with irradiation would undoubtedly affect the egg industry, too. "It's a new concept, and the public has to get used to it," Josephson says, adding that many people still associate irradiation with atom bombs and don't consider the many beneficial medical applications that radiation has for treating and diagnosing cancer and other diseases. There are also some organizations that oppose any use of radiation, "so they muddy the water by spreading half-truths or untruths," he says.

Another way to rid shell eggs of SE is to pasteurize the egg in the shell, and several research groups are exploring the best way to do this. PREEMPT—the new spray-on blend of twenty-nine live, harmless bacteria that crowd out the growth of pathogenic *Salmonella* in a chicken's intestine—has the potential of drastically reducing the number of SE-infected laying hens. Over time, that could translate into fewer SE-infected eggs, or perhaps the elimination of SE.

In the meantime, some states have restricted eating establishments from serving undercooked eggs or using raw eggs in menu items without the customer's knowledge. And many commercial and institutional food-service establishments have begun using pasteurized egg products instead of pooled raw eggs to reduce the risk of *Salmonella* infection outbreaks. According to the CDC, other efforts are under way to reduce the SE threat on a broader scale. These include:

- Many states now require refrigeration of eggs from the producer to the consumer.
- The egg industry has voluntarily instituted some quality assurance and sanitation measures.
- The USDA has announced a mandatory program to test the breeder flocks that produce egg-laying chickens to make sure they are SE-free.
- When an egg-laying flock is implicated as the most likely source of contaminated eggs in an outbreak of SE infections, the USDA will require that flock to be tested.
- The FDA is preparing regulations to monitor infection in laying hens should the USDA program prove insufficient.

- The FDA has issued specific guidelines for handling eggs in retail establishments.

Despite all these measures, egg lovers should not let down their guard, especially if their age or health status raises their risk for contracting a food-borne disease. It is impossible to tell by looking at an egg or tasting it whether it contains *Salmonella*.

The importance of treating raw and undercooked eggs and egg products as potentially hazardous was underscored by an exposé on the NBC television news program *Dateline* in April 1998. Food-safety experts say that fresh shell eggs should be eaten within three to five weeks of purchase. But a *Dateline* hidden camera showed eggs nearly a month old being rewashed, repacked, and redated at an egg farm in Ohio that produces about two million eggs a day and distributes them to twenty states. According to the *Dateline* report, these old eggs were routinely placed with new eggs (in the same cartons), given a USDA grade-A seal, and sold to the public as fresh. According to its survey of all fifty states, *Dateline* found that only Wisconsin "flatly prohibits" rewashing and redating eggs and also requires an expiration date of no more than thirty days on every carton.

At the time of the report, repacking and redating eggs up to twenty-nine days old was perfectly legal, and according to United Egg Producer, an industry group, the practice did not pose a health risk to consumers. ". . . There is no reason to believe that the reprocessing of eggs has any effect on the quality, freshness, or safety of the product," United Egg Producer said in a statement to *Dateline.*

Scientists and medical experts are not so sure. Research has shown that a single *Salmonella enteritidis* bacterium in a shell egg can multiply to 100,000 in a month's time if the egg is kept at 50 degrees. If the egg is kept at room temperature, the growth rate could be one hundred times faster. *Dateline* reported that the temperature inside one cooler at the Ohio plant read 53 degrees.

A couple of weeks after the report aired, Agriculture Secretary Dan Glickman announced a prohibition on the repackaging of eggs packed under USDA's voluntary grading program. "We want to make sure consumers can rely on the USDA shield on egg cartons as a symbol that they are purchasing high-quality eggs," Glickman said in a news release. About one-third of all shell eggs sold to consumers are graded by the USDA. Among other things, the USDA shield means:

- The eggs were packed under the supervision of a USDA grader who checks samples of the eggs before they are shipped to the retailer; and
- The egg packer meets USDA's sanitary standards.

Egg packers who do not use the USDA grading service may put terms such as "Grade A" on their egg cartons, the press release states, but they may not use the USDA shield.

Several weeks after Glickman's announcement, the USDA and FDA asked for public comments on a new egg-safety plan covering such issues as preventing the introduction of SE in laying chickens, and changing egg processing, handling, and storage procedures to further enhance egg safety. The agencies are also proposing to:

- Control the temperature maintained by vehicles transporting shell eggs to market;
- Require egg cartons to carry labels with safe food-handling instructions;
- Maintain cool temperatures in supermarkets and food-service establishments;
- Institute a nationwide surveillance program to track the spread of SE infection among laying flocks; and
- Launch public-private partnerships to develop a national food-safety education and training campaign for food-service workers and consumers.

Egg-Safety Guidelines

You can use eggs more safely by following these guidelines from the CDC, the American Egg Board, the USDA's Food Safety and Inspection Service, and the Center for Science in the Public Interest:

- Buy refrigerated grade AA or A eggs with clean, uncracked shells. (There is no relationship between shell color and egg safety.)
- Only buy eggs if they are displayed in a refrigerated case.
- As soon as you get home from the market, store

eggs in their original carton in the coldest part of the refrigerator. (Do not wash eggs before storing or using them. Washing is a routine part of commercial egg processing, and rewashing is unnecessary.)

- If any eggs accidentally crack after purchase, break those eggs into a clean container, cover tightly, refrigerate, and use within two days.
- Don't eat raw eggs. This includes traditional recipes for homemade ice cream and eggnog. If the recipe calls for an adequately cooked custard, however, the eggs should be safe. (Commercially manufactured ice cream and eggnog are made with pasteurized eggs.)
- Use only pasteurized eggs for Caesar salad dressing, homemade mayonnaise, eggnog, and meringue.
- Use raw shell eggs within three to five weeks of purchase. Use hard-boiled eggs (in the shell or peeled) within one week after cooking. Use leftover yolks and whites within four days of removing them from the shell, and only if they were refrigerated after being removed from the shell.
- When cooking eggs, let the yolk thicken to the point where it is no longer runny. Eggs fried "over easy" are probably safer than eggs fried "sunny-side up," because the yolk is likely to be better cooked.
- Don't let cooked eggs sit at room temperature for more than two hours, even if they are in a warming tray. Eggs are safest when eaten as soon

as they are cooked. (If you hide hard-cooked eggs for an Easter egg hunt, either follow the two-hour rule or do not eat the eggs.)

- When refrigerating a large amount of food containing hot eggs, divide it into several shallow containers so it will cool quickly.
- Wash hands, utensils, blenders or other kitchen appliances, and work areas with hot, soapy water before and after they come in contact with eggs or egg-containing foods.
- At a restaurant, find out whether the cook uses pooled shell eggs or pasteurized egg products to make scrambled eggs, egg-based sauces, omelets, French toast, and sandwiches before you order them.

Milk Safety

Before pasteurization of milk was introduced in the early part of the twentieth century, milk-borne illnesses were not unusual. Tuberculosis, brucellosis, typhoid fever, polio, and dysentery were among the diseases transmitted by raw milk. Today, almost all milk sold in the United States is pasteurized (heated to a temperature that will kill unwanted bacteria), and illness traced to pasteurized milk or the cheeses made from it are extraordinarily rare. "It's reasonable to trust that if you buy pasteurized milk, your risk of contracting a food-borne infection is next to nothing," says Cornell food scientist David Bandler.

When an outbreak stemming from pasteurized

dairy products does occur, it usually is traced to a correctable problem with the pasteurization process, such as a crack in a pipe or a mechanical failure, or contamination of the product during transit. Bandler says that certain pathogenic bacteria originate from the environment the milk-producing cow lives in and may be carried from farms to dairy plants by vehicles that pick up the milk. Although pasteurized milk can potentially become recontaminated once the carton is open, it will probably spoil before it becomes a major health threat. As mentioned at the beginning of this chapter, keeping milk as cold as possible is the key to lengthening its shelf life.

Pasteurization involves heating raw milk to at least 145°F for thirty minutes or more, or to at least 161°F for fifteen seconds, and then rapidly cooling it down to 40°F or lower. This process kills disease-causing microorganisms. While pasteurization makes milk safe, raw (unpasteurized) milk and the cheeses made from it pose an appreciable health risk. Among the pathogens that may infect unpasteurized milk and dairy products are *Bacillus cereus, Listeria monocytogenes, Yersinia enterocolitica, Salmonella, Campylobacter jejuni, E. coli* O157:H7, and the other Shiga-toxin-producing *E. coli*.

Canada bans the sale of raw milk for human consumption, and raw milk sales across state lines are prohibited in the United States. According to the FDA, it is illegal in twenty-two states as well as in the District of Columbia and Puerto Rico to sell raw cow or goat's milk to consumers. Of the remaining states, most allow unpasteurized milk to be sold only on farms. As of 1995, the most recent year for which FDA data were available, selling raw milk in grocery stores was legal in Arizona, Arkansas, California, Connecticut, Idaho, Maine, New Hampshire, New Mexico, Oregon, South Carolina, and Washington. Six states (Arkansas, California, Idaho, Maine, New Mexico, and Oregon) also allow restaurants to sell raw milk. As mentioned above, even in states where raw milk sales are legal, the vast majority of milk sold is pasteurized and labeled as such. In 1997 the Washington Department of Agriculture announced a new rule requiring sellers of raw milk in that state to label containers: "WARNING: This product has not been pasteurized and may contain harmful bacteria. Pregnant women, children, the elderly and persons with lowered resistance to disease have the highest risk of harm from use of this product." According to *FARMComm,* an on-line news and information service directed at rural and farm communities, Washington's label is intended to ensure that consumers are making an informed choice, and it was implemented primarily in response to the emergence of new and more dangerous bacteria. Washington reported twenty-three cases of illness between 1990 and 1995 that were linked to unpasteurized milk. In 1997 eight children became ill after drinking raw milk during a dairy tour in King County, Washington.

According to Cornell's David Bandler, some people prefer raw milk because they mistakenly believe that unprocessed foods are, by definition, always safer and more wholesome than processed foods. But Bandler says, "I think everybody in the scientific community will agree that drinking raw milk is a bad idea." The scientific and medical communities are most concerned about city-dwellers who buy raw milk at farmers' markets, he says.

Cheese Safety

Perhaps the most common dairy product recalled because of bacterial contamination is a Mexican-style cheese known as queso fresco, or fresh cheese, which Bandler calls one of the few dangerous cheeses. Unlike cottage cheese and many other dairy products, queso fresco is not protected by the acidity that is needed to help curtail bacterial growth. When queso fresco is made from unpasteurized milk, the risk of contamination is even higher. Every attempt to acidify queso fresco results in reduced sales, Bandler says, adding that this cheese is "kind of like plain milk; it has no built-in protection from microbial growth." Pasteurization, protection in a properly sealed container, and refrigeration make milk safe. "But this product, even when it's made from pasteurized milk, is open to the air and the hands of people who manufacture it," he says.

The problem got so bad in California that in early 1998 the state's dairy industry teamed up with state and local health officials to launch a campaign to educate consumers about the potential health risks associated with consumption of these cheeses made by unlicensed manufacturers. About 46 million pounds of licensed queso fresco are produced in California each year, representing about 4 percent of the California cheese market. However, state officials believe that much more queso fresco is made and sold by unlicensed vendors.

According to Kimberly Belshe, director of the California Department of Health Services, *Salmonella* and *Listeria* are the two most frequent bugs found in illegally produced queso fresco.

Bacteria are not the only health hazards that can be transmitted through dairy products. Molds—primarily *Aspergillus, Fusarium,* and *Penicillium*—can grow in milk and cheese. Just because cheese gets moldy, however, doesn't mean it is completely inedible. You can cut mold off hard cheese, such as cheddar, and safely eat the remainder, particularly if mold is blue and similar to what you see in blue cheese or Stilton. But if the mold is pink, white, or multicolored, it may produce toxins, which are a potential health hazard. Bandler cautions against attempting to salvage soft cheeses, such as cream cheese and cottage cheese, that have become a little moldy. "The moisture content of these products is such that if mold is growing on it, there are probably some other things happening within it, too," he says. "Basically, the higher the moisture content of the cheese, the more likely it is to support growth of a variety of organisms, and without a laboratory, you can't tell. So the best thing to do is chuck it."

8

SAFETY ON TAP:

DRINKING WATER

At the turn of the century, America's drinking water was often hazardous to human health, transmitting a host of diseases, inducing such scourges as cholera, typhoid fever, and amebic dysentery. The incidence of these and other waterborne diseases has been reduced dramatically since the advent of water treatment in the early 1900s. Today, virtually all public water and most private supplies are disinfected, usually by chlorine. Water chlorination is considered one of the greatest public-health triumphs of the twentieth century.

But chlorination is not always enough to make drinking water safe. Water-treatment plants that draw on surface water sources, such as lakes and rivers, are generally required to use a "multibarrier approach"—disinfection plus filtration or other technologies to remove microorganisms. Public water supplies drawing upon groundwater for the most part use only disinfection. The Safe Drinking Water Act is designed to ensure that the levels of known contaminants in public water systems are too low to present a significant public-health risk. Enforced by the Environmental Protection Agency (EPA), the Act is updated periodically to reflect evolving scientific knowledge about waterborne contaminants. Although advocacy groups, such as the Environmental Working Group, contend that some 50 million Americans consume potentially disease-causing microorganisms and chemicals in their drinking water, the United States' multi-billion-dollar water-treatment system is considered by many scientists one of the best, if not the best, in the world.

Yet, despite continuous monitoring for contaminants by water utilities, problems occasionally occur. Water is a universal solvent, and no water system is capable of removing—or is designed to remove—100 percent of everything. Water treatment and distribution systems are geared toward reducing contamination to the point where any remaining pathogens are too few and far between to cause disease. Every so often, however, a water system temporarily fails to comply with the safety standards set forth in the Federal Safe Drinking Water Act. Contamination levels may rise because of equipment malfunction, human error, severe weather, industrial pollution, or other factors. In 1993 and 1994, there were 30 reported waterborne-disease outbreaks, 23 of which were associated with public drinking water supplies, and 7 with private wells, according to the EPA. Many of these outbreaks have been linked to contamination by bacteria or viruses, probably from human or animal wastes. According to researchers at the University of Wisconsin, a class of Norwalk viruses called caliciviruses has also been traced to recent waterborne outbreaks in the United States. The bacterium *E. coli* O157:H7 has caused at least two waterborne outbreaks that affected more than 300 people. In North and South America, the parasitic protozoan *Cyclospora* has caused serious health problems for people relying on untreated drinking water.

According to the EPA, of the more than 55,000 community water systems in the United States, 4,769 or 8.6 percent reported a violation of one or more drinking water health standards in 1996, the most recent year for which data are available. A spokeswoman at the Safe Drinking Water Hotline points out that these violations do not necessarily mean there was contamination per se, or that public health was in jeopardy. A violation could be failing to test the water for a certain contaminant in a timely manner, she says.

Water systems must continuously monitor their supplies because contamination can come from a multitude of sources. "Suburban sprawl has encroached upon once-pristine watersheds, bringing with it all of the byproducts of our modern lifestyle," states a 1997 EPA publication, *Water on Tap: A Consumer's Guide to the Nation's Drinking Water.*

"Chemicals can migrate from disposal sites and contaminate sources of drinking water. Animal wastes and pesticides may be carried to lakes and streams by rainfall runoff or snowmelt. Human wastes may be discharged to receiving waters that ultimately flow to water bodies used for drinking water. Coliform bacteria from human and animal wastes may be found in drinking water if the water is not properly treated or disinfected." (*Water on Tap* is on line at www.epa.gov/OGWDW/wot/wot.html. To obtain a printed version, call the Safe Drinking Water Hotline at (800) 426-4791 and ask for document #EPA 815-K-97-002. Allow four to six weeks for delivery.)

When violations of the Safe Drinking Water Act occur, the water system is required to issue a "boil-water notice" that remains in effect until the problem is rectified. Over the past few years, more than 725 communities, including New York City and the District of Columbia, have issued boil-water notices affecting a total of about 10 million people.

An increasing number of boil-water notices in recent years may partly explain why a 1997 national survey indicated that three-quarters of the 1,003 Americans questioned had concerns about their household water supplies. According to the survey, commissioned by the Water Quality Association and conducted by an independent opinion-research company, the number of people who installed filters or other point-of-use water treatment devices rose 5 percent between 1995 and 1997.

A desire to improve the taste of water is certainly a valid reason to filter your tap water, but paranoia about waterborne pathogens is not, many experts say. In fact, unless you have AIDS, HIV, or another immune disorder, there are no general health reasons to avoid drinking water straight from the tap. For healthy individuals, says Richard Levinson, M.D., D.P.A., of the American

Public Health Association, "I think the danger has been blown up out of all reasonable proportion."

Because chemical disinfection has virtually eliminated bacterial and viral threats from our water supply, the biological threats of most concern are protozoa, especially *Cryptosporidium parvum* and *Giardia lamblia*. These parasites generally originate from mammalian feces in water and are extremely resistant to inactivation by chlorine. According to a report in the CDC journal *Emerging Infectious Diseases, Giardia* and *Cryptosporidium* were the leading causes of 129 disease outbreaks in the United States from 1991 through 1994 that were traced to drinking, or swimming in, contaminated water.

Cryptosporidium

The infectious form of *Cryptosporidium* (or *Crypto*) is its chlorine-resistant egg-shaped capsules, or oocysts, each of which contains four parasites. Infection with these oocysts can lead to cryptosporidiosis, a disease characterized by watery diarrhea and cramps, and possibly weight loss, nausea, vomiting, and fever. The disease is self-limiting for most healthy individuals, with symptoms lasting approximately one to two weeks. People with damaged immune systems are at a higher risk for severe, possibly fatal complications of this disease, for which there is no known treatment.

People may become infected by drinking contaminated water, but it is also possible to ingest *Crypto* while swimming in a pool, river, lake, or pond. "If you look hard enough for *Crypto* in most watersheds, you will find it," says David Battigelli, Ph.D., adjunct professor of preventive medicine at the University of Wisconsin at Madison and chief of environmental virology at the Wisconsin State Laboratory of Hygiene. He says there is a general feeling among scientists that *Crypto* is entering surface water supplies from cows living in concentrated farming operations. These operations are typically situated near large bodies of water. "But try as we might," Battigelli says, "we've never been able to identify *Crypto* causing a human outbreak as coming from a particular batch of cows."

Crypto concentrations can also be reduced by membrane filtration, reverse osmosis, ozonation, and other technologies, but relatively few water systems use these methods. And they may not need to. According to biologist Walter Faber, Ph.D., a professor at Manhattan College in the Bronx and editor of a monthly newsletter, "*Cryptosporidium* Capsule," *Crypto* concentrations in U.S. drinking water supplies are ordinarily too low to produce human disease.

Under highly unusual circumstances, there can be enough *Crypto* in a water source to overwhelm a treatment facility's capacity to bring it down to innocuous levels. When inadequate water treatment in Milwaukee, Wisconsin, coincided with heavy rainfall and runoff at the treatment plant in 1993, some 400,000 residents contracted cryptosporidiosis from drinking water, and sixty-seven died. According to the EPA, it was the biggest waterborne disease outbreak ever recorded in this country. Since that time, many water systems have added filtration equipment designed to significantly reduce *Crypto* concentrations, and a handful of water utilities have added ozonation technology. One recent study noted that the number of cryptosporidiosis cases in the United States has declined since 1996.

Currently, the EPA has no limit, or "standard," for *Crypto* in drinking water. However, the EPA has directed public water plants that serve populations of 100,000 or greater to assess the distribution of *Crypto* in surface waters throughout the United States to see how widespread the problem is. As part of this study, these water plants are monitoring how effective their treatment processes are in removing *Crypto*. If *Crypto* is found to be a major problem in treated water, the EPA will set a standard limiting *Crypto* concentrations in drinking water.

Various experts have found the infectious dose of *Crypto* to be 30 to 123 oocysts. Faber says the 123-oocyst dose was determined by studies conducted on extremely healthy volunteers at the University of Texas at Houston. Most of the test subjects who ingested that much *Crypto* had diarrhea for three weeks, he notes. According to Faber, routine surveillance for *Crypto* in treated water supplies has found no more than 1 oocyst per 100 liters.

While the overall risk of contracting cryptosporidiosis from drinking water is very small, even for AIDS patients, the public-health community advises those with weakened immune systems to boil their drinking water or to filter it with a "one-micron" filter—just in case. Bringing water to a rolling boil for one minute kills *Crypto* oocysts.

Giardia

Like *Crypto, Giardia* can also be killed through boiling or by chlorination but only after an unusually long exposure to the disinfectant. According to the CDC, *Giardia* has been the cause of nearly all reported outbreaks of waterborne parasitic diseases in recent years; it only takes the ingestion of one *Giardia* protozoan to cause symptoms. In each of the outbreaks, water chlorination was adequate to make outbreaks of bacterial diseases unlikely. However, lack of an intact system capable of filtering *Giardia* cysts, problems with the water-distribution system, or mechanical deficiencies have allowed drinking water to become a vehicle for giardiasis. Studies have shown that many wild and domestic animals, including 90 percent of dogs, carry this organism, so washing hands well after touching or walking your pets or visiting a petting zoo is extremely important.

In humans, *Giardia* infection usually causes diarrhea or foul-smelling stools, cramping, excessive gas or bloating, and fatigue but no fever. Symptoms tend to last three to twenty days, but may wax and wane for several weeks thereafter. The EPA has a zero level of tolerance for *Giardia* in drinking water. Infection with *Giardia* is a frequent problem in those returning from trips to developing countries; travelers with symptoms such as those outlined above may wish to suggest *Giardia* screening to their physicians. Giardiasis is usually treated with a medication called metronidazole (Flagyl). Children under age five may be treated with furazolidone (Furoxone). According to the American Academy of Family Physicians, the latter has fewer side effects and comes in a liquid form, but it shouldn't be given to babies younger than one month old. Because giardiasis is highly infectious and can be spread from person to person, doctors may wish to treat the whole family simultaneously.

Disinfection By-Products

Rebecca, a mother of a kindergartner and a toddler, says she installed a water filter on her kitchen sink several years ago because the pipes in her home are old and may contain lead, and also because she is skeptical of the government's ability to keep her water supply safe. In particular, Rebecca says she worries about potentially cancer-causing disinfection by-products, or DBPs, which, according to one controversial study, may elevate the risk of miscarriage. DBPs are chemicals formed when chlorine or any disinfectant, including ozone, reacts with organic matter, such as decaying leaves and soil. Organic matter is far more common in surface water, such as reservoirs and rivers, than it is in groundwater that gets filtered through hundreds of feet of geologic formations.

The EPA is tightening the standards for the most common class of DBPs, trihalomethanes, from 100 parts per billion (ppb) to 80 ppb and may further reduce it to 40 ppb at some point in the future, according to Joseph Harrison, technical director for the Water Quality Institute, a trade organization of filtration and treatment equipment manufacturers. An example of a trihalomethane is chloroform. Some researchers are urging the EPA to reduce DBP standards further even though almost thirty years of scientific scrutiny has produced only "hints" that these chemicals in drinking water may increase the risk of colorectal and bladder cancer, according to *Health News,* a consumer publication produced by the *New England Journal of Medicine.*

Keith Christman, director of disinfection for the Chlorine Chemistry Council, a trade group, notes that the EPA was slated to finalize a regulation in November 1998 requiring selected water systems to install "enhanced coagulation" technology. Enhanced coagulation is a gravity process designed to precipitate out most of the organic matter before disinfection, thus reducing the DBPs and making chlorination more effective. A few water systems already have enhanced coagulation, Christman says, and the rest of the water systems affected by the regulation—primarily those that use surface water sources—will be given a few years to get onboard. "The water-treatment industry is very capital intensive, very expensive," he says. "It takes some time for them to adapt."

Getting Information About Your Water Supply

Unless you work for your local water company or you test your private well, you probably have no idea whether your household water contains DBPs, *Giardia lamblia, Cryptosporidium parvum,* or other microbes or chemicals, and if so, at what concentrations. That will change in 1999 when public water companies issue the first annual "Consumer Confidence Report." This document, which some water companies are expected to rename "Water Quality Report," was mandated in 1996 by amendments to the Safe Drinking Water Act. Larger water companies will mail their reports to customers' homes; systems serving 10,000 customers or fewer may make the report public through other means.

At the very least, the report must contain:

- information about where your drinking water comes from;
- results of monitoring that the system performed during the year; and
- information on health concerns associated with violations that occurred during the year.

The 1996 amendments further require the EPA to compile and summarize state water-quality reports into an annual report on the condition of the nation's drinking water. Until those reports become available, you can find out how your household water supply stacks up to state and federal standards by asking your water utility department for a copy of the "Municipal Drinking Water Contaminant Analysis Report." If you use a private well, you can ask your local health department for a list of the typical well-water contaminants in your area, advises NSF International, a not-for-profit organization that, among other things, certifies home water-treatment devices. A third option—one that public-health experts say is generally unnecessary—is to have your tap water tested by a certified water specialist or laboratory. Many certified water specialists are also in the business of selling water filters. Your local health department should be able to provide you with a list of labs certified by your state to test water. If you have specific questions, call the EPA's Safe Drinking Water Hotline.

Home Water Filters

Nearly one-third of respondents to the Water Quality Association (WQA) survey said they filtered or otherwise treated their water at home, even though there is no general health reason to do so—unless you have AIDS or another condition that suppresses the immune system. The WQA's Joseph Harrison says that healthy people who install home water-treatment devices "often like to have the taste improved or want peace of mind that things that the city can't afford to remove can be taken out at the home." Indeed, a homeowner can buy a "nanofiltration" system equipped with a sophisticated membrane that can catch particles as small as viruses. Particle size is measured in micrometers (or microns), which are equal to one-millionth of a meter. The diameter of a human hair is about 50 micrometers; a virus is about 3/100ths to 10/100ths of a micrometer, and a bacterium measures about 1 micrometer. *Cryptosporidium parvum* oocysts measure 4 to 6 micrometers.

There are two general categories of home water-treatment systems—point-of-use (POU) and point-of-entry (POE). According to NSF International literature, POU styles include:

- Pour-through: Water drips through a filter by gravity into a pitcher that is usually stored in the refrigerator.
- Personal water bottle: A filter is integrated into a push-pull cap or straw.
- Faucet mount: The unit is mounted on the existing kitchen-sink faucet, usually replacing the aerator or installed above the aerator. A diverter is usually used to direct water through the system when filtered drinking water is desired.
- Countertop connected to sink faucet: This unit is usually placed on a counter and is connected by tubing to an existing kitchen-sink faucet.
- Plumbed-in: This unit is generally installed under the sink and permanently connected to the water pipe. The unit directs filtered water through an existing faucet or auxiliary faucet mounted next to the kitchen sink. A licensed plumber may be needed to install it.

POE systems typically treat most or all of the water entering a house.

Home water-treatment devices can range in price from $25 or less for a pitcher with an activated charcoal filter, to $500 or $1,000 for a plumbed-in device, to $2,000 for a POE system. Water-treatment experts emphasize that all systems require periodic maintenance in order to operate effectively. For example, if ultraviolet lights aren't cleaned periodically, they will get covered in moss and slime, which will render a UV water-treatment unit ineffective.

Ultraviolet light, which treats water by killing microorganisms, is one of several treatment technologies available to homeowners. Others include:

- Adsorption—the physical process that occurs when liquids, gases, or dissolved or suspended matter adhere to the surface or pores of an adsorbent medium, such as carbon or charcoal.
- Filtration—the process by which particles are separated from water by passing through a semipermeable material.
- Reverse osmosis—a water purification system that reverses, by the application of pressure, the flow of water from a more concentrated solution to a more dilute solution through a semipermeable membrane.
- Distillers—a process that consists of vaporizing water and condensing it back to a liquid state.

Different systems remove different things from water, and many remove multiple contaminants. Dean O. Cliver, Ph.D., of the University of California at Davis, says consumers should be "fairly sure" of what needs to be removed from their tap water before investing in a water-treatment unit, whose purchase and maintenance costs can be substantial. "It is important not to pay to treat water in ways that are not necessary," he says.

If you are shopping for a water-purification unit, look for a mark on the label indicating that it has been independently tested to confirm that the unit does what the manufacturer claims it does. The WQA and NFS International are two of several organizations that perform this testing service. The WQA offers its "Gold

Seal" to products it certifies, and the NFS certification appears as a blue-and-white or black-and-white circle with the letters "NFS" inside. The label should also specify what the unit is certified to reduce or remove from water, including "VOCs" (volatile and nonvolatile organic chemicals, usually from pesticides), "trihalomethanes" (and other disinfection by-products), "cysts" (*Cryptosporidium, Giardia,* and other parasites), lead, and radon. (For more information on radon removal from water, call the Consumer Research Council's Radon FIX-IT Hotline, (800) 644-6999.) Many units also remove chlorine, sodium, and other substances that can affect water aesthetically, although some units, such as water softeners, actually add sodium, which has health implications for people with heart disease or high blood pressure.

NSF International has an interactive Web site through which consumers can obtain a list of NSF-certified units that are effective in removing the particular contaminants they are concerned about. For price and purchase information, or if you do not have Internet access, you may call NSF at (800) 673-8010 and request a copy of the *Consumer Book on Drinking Water Treatment Units.* A wealth of information on drinking water is also available on the Water Quality Association's Web site: www.wqa.org and on the American Water Works Association Web site: http://www.awwa.org/asp/.

Bottled Water

Bottled water is generally a tastier, albeit more expensive, option to tap water, but not necessarily a safer one from a microbiological standpoint. Like tap water, most bottled water contains some level of carbon, which will support the growth of bacteria unless the water is bottled under completely sterile conditions—which most bottled water is not, says Battigelli. "It's unlikely you'll have water with no viable microorganisms in it," the researcher says. "The organisms may not be pathogenic, but bottled water is seldom sterile."

Battigelli, of the Wisconsin State Laboratory of Hygiene, urges consumers to read the fine print on bottled-water labels to find out where the product came from. For example, a company can draw their product from a municipal water system and may or may not put the water through additional treatment before bottling. Biologist Walter Faber recommends boiled rather than bottled water for anyone with immune system damage. For immune-compromised individuals who prefer the taste or convenience of bottled water, Battigelli recommends products that are "micron filtered" (passed through a one-micrometer filter) and ozonated. A one-micron filter is capable of removing all or most *Crypto* oocysts.

In general, Battigelli says, drinkers of bottled water, regardless of their immune status, should consume the product as soon as possible after purchase. Storing it in an air-raid shelter for three years may not make the water unsafe, he says, but it probably will render it foul tasting.

The Food and Drug Administration, the states, and the bottled-water industry all regulate bottled water. For more information, contact the International Bottled Water Association, 1700 Diagonal Road Suite 650, Alexandria, VA 22314; (703) 683-5213; or visit its Web site: http://www.bottledwater.org/.

Looking Ahead

Battigelli is part of a research team investigating ways of identifying and eliminating new pathogens that may constitute future threats to drinking water safety. "If we review waterborne outbreaks of infectious disease in the United States, we find that the culprit was identified in only about half the cases," says Battigelli's colleague, civil engineering Professor Greg Harrington of the University of Wisconsin at Madison. "There are numerous microorganisms, but detection methods are available for only a small fraction."

To help accomplish their mission, Harrington, Battigelli, and a third scientist, Jon Standridge, are building a miniature water-treatment plant about the size of a tractor-trailer. Water pumped into the model plant will be spiked with cocktails of about a dozen strains of bacteria and parasites and then treated in a variety of ways. The approach will enable the researchers to evaluate the ability of different treatment methods to remove the bugs. The two-year research project is funded by a $250,000 grant from the EPA and the American Water Works Association Research Foundation, an international nonprofit group devoted to drinking water quality.

One of the pathogens in the cocktail is *Microsporidium.* This parasite appears to cause human illness, particularly in immune-suppressed individuals, but little is known about its fate, ecology, or distribution in the environment, Battigelli says. By contrast, a great deal is known about *Crypto,* which also will be put in the cocktail. "By testing both under all of our water-treatment conditions, we might be able to get an idea of how removable *Microsporidium* is compared to *Crypto,*" he says. *Microsporidium* was selected deliberately because it is about one-fifth the size of *Crypto,* which theoretically makes it harder to remove by traditional filtration processes; traditional sand-filtration methods are generally more effective with larger particles. The team will also be studying viruses one hundred times smaller than *Crypto,* viruses that could more easily pass through some filtration systems. New water-treatment technologies they will be looking at include dissolved air flotation and microfiltration.

Harrington says he believes the *Crypto* threat has made the public more willing to pay for better technologies. According to a press release from the University of Wisconsin, the city of Kenosha, Wisconsin, is installing a microfiltration system for city water, and Milwaukee is upgrading its filters and installing ozonation. In 1999, Cambridge, Massachusetts, plans to break ground on a new $50 million water-treatment plant using both activated carbon and ozone, according to a recent report on the National Public Radio program *Living on*

Earth. It is the largest public-works project in the city's history, the report noted.

Harrington emphasizes that U.S. water-treatment standards are very good, and that drinking water is generally quite safe. "We could look at thousands of water samples before finding anything to worry about," he says. "But the consequences of failing to remove pathogens can be very large. We want to ensure there is minimal risk of public exposure."

Drinking Water Safety Tips

- In general, tap water from public water supplies is safe to drink. When contaminants exceed safety limits, water customers are notified to boil their drinking water until the problem is resolved.
- If you have AIDS or another condition that suppresses the immune system, boil your drinking water or filter it with a one-micron filter.
- If you opt to buy a home water-treatment unit to improve the quality or aesthetics of your drinking water, look for a certification seal or logo on the box indicating that the device has been independently tested. Be sure to maintain the device (i.e., change the filter at recommended intervals) according to the manufacturer's recommendations so it will operate at peak efficiency.
- If you want to know what is in your household

water supply, ask your water utility department for a copy of the "Municipal Drinking Water Contaminant Analysis Report." If you use a private well, ask your local health department for a list of the typical well-water contaminants in your area.

- If you have questions about your tap water, call the EPA's Safe Drinking Water Hotline at (800) 426-4791.
- If you have questions about radon in your water supply, call the Consumer Research Council's Radon FIX-IT Hotline, (800) 644-6999.
- Keep bottled water refrigerated and drink it as soon as possible after purchase.

PART 3

FOOD-SAFETY GUIDELINES

9

REDUCING YOUR RISK FOR FOOD-BORNE ILLNESS AT HOME

A freshman-year science experiment forever changed Annie Condit's hand-washing habits. She was taking a microbiology class at New York's Oneonta State University College in 1979. At the beginning of the semester, each student received a "mystery microbe," which they were assigned to identify by the end of the course. The professor insisted on creativity. "I was playing with it, incubating it to see how it would grow and multiply," Annie recalls. "Then I got the idea of comparing its growth with the growth of everyday microbes that are found on people's hands."

The next morning, she awoke around seven o'clock and washed up as usual—but refrained from washing her hands again until four o'clock that afternoon. "I wanted to see what I'd pick up during an average day in college without washing my hands," she remembers. "I touched doorknobs, toilet seats, desks, books and magazines, money, public phones, my car, my hair, my face, a slice of pizza for lunch, my clothes, my shoelaces, a typewriter, whatever."

When she returned to the lab later that day, Annie lined up four petri dishes containing a sanitized gel, or "growth medium," on which any bacteria that were introduced could multiply unencumbered. Annie pressed her unwashed fingers onto the growth medium of petri dish 1. Then she rinsed her hands with cold water, dried them with a paper towel, and pressed her fingers onto the gel of dish 2. Before inoculating dish 3, she washed with soap and warm water for twenty or thirty seconds before drying her hands with a paper towel. For dish 4, she really went to town. She had obtained some antiseptic soap and a scrub brush from a hospital surgical suite, the same kit surgeons use for washing up before operating. Annie used the kit the same way a surgeon would, scrubbing between each digit and under the nails; she even scrubbed her wrists and forearms. Annie placed all four petri dishes in an incubator, where they remained for forty-eight hours.

Upon inspecting the results with her naked eye and under a microscope, Annie was astonished. There were hundreds of thousands of colonies of bacteria in dishes 1 *and* 2, leading Annie to conclude that simply rinsing with water did nothing to remove the day's worth of microorganisms that had been teeming on her skin. By comparison, there was almost nothing growing in dish 3, and dish 4 was clean. "There was maybe a five-colony difference between dishes 3 and 4," Annie recalls. She photographed her results, wrote up her findings, and displayed them on a poster in the lab for other students to see. Since then, she says, "I've been a hand-washing freak."

Annie got an A in microbiology, and went on to attain a master's degree in nutrition from New York University.

As graphically demonstrated by Annie's simple experiment—and by sophisticated studies conducted by professional scientists—washing your hands with soap and warm water is one of the most powerful weapons you have to fight the spread of food-borne illnesses (and many other infectious diseases). Controlling the temperature of your food is another critical measure. Food-safety specialists continually stress the importance of cooking meat and poultry until they reach appropriate temperatures and keeping hot foods hot and cold foods cold until they are eaten. Another important way to reduce your food-borne-illness risk is to routinely sanitize your utensils, cutting board, table, and counters.

Some or all of this advice may seem obvious to you. Yet research shows that very few people—even highly educated individuals—are putting their food-safety knowledge into action. For example, researchers have reported that 53 percent of consumers eat foods containing raw eggs, 23 percent eat undercooked hamburgers, 17 percent eat raw clams and oysters, and 26 percent fail to wash their cutting boards after using them for raw animal products.

One recent study documented serious food-safety

violations in a vast majority of households whose food-handling practices were evaluated on a volunteer basis. Evaluators from Audits International of Highland Park, Illinois, observed meal preparation, service, post-meal cleanup, and leftover storage in 106 households in 81 cities across the United States and Canada. According to the February 1998 report in the journal *Food Technology,* participants were better educated than the average American and were therefore more likely to perform better than randomly selected households. Yet 96 percent of households had at least one critical food-safety violation, which, by itself, could potentially lead to a food-borne infection. Critical violations include cross-contamination of food, neglected hand washing, undercooking of foods, food handled by people with cold or flulike symptoms, and improper cooling of leftovers. The average number of critical violations per household was 2.8. The auditors used standards recommended in the FDA's 1997 Food Code. Local, state, and federal regulators use the FDA Food Code as a model to help develop or update their own food-safety rules and to be consistent with national food regulatory policy.

"We can complain about processing facilities, distribution systems, supermarkets, and restaurants, but we must also take responsibility for ourselves," says Richard W. Daniels, president of Audits International and the author of the *Food Technology* article. "Proper preparation at home is the last step, and in some cases, the last chance we have to protect ourselves."

Toward that end, this chapter offers a practical, step-by-step guide designed to help you hone your current food-safety skills and adopt new ones for buying, preparing, cooking, and storing food, reusing leftovers, and cleaning up. The suggestions were culled from a variety of sources, including Rutgers Cooperative Extension, the *Tufts University Health and Nutrition Letter,* the *U.C. Berkeley Wellness Letter,* the U.S. Centers for Disease Control and Prevention, the Food and Drug Administration, the U.S. Department of Agriculture's Food Safety and Inspection Service, and the USDA's Meat and Poultry Hotline.

Hand Washing

Do you always wash up after using the bathroom? Almost one third of consumers (32 percent) do not, according to a 7,000-person survey sponsored by the American Society for Microbiology in 1996. And although 81 percent of survey respondents said they wash their hands before handling or eating food, observational studies draw a different picture. For instance, one recent study found that only 21 percent of males and 58 percent of females washed properly after using a public toilet. In 1992 the *New England Journal of Medicine* published a study suggesting that even health-care workers failed to follow hand-washing protocols 60 percent of the time.

Many food-borne pathogens such as *Shigella, Escherichia coli, Salmonella,* and hepatitis A virus can hitch a ride on hands. These bugs may also be spread through direct contact with the stool of an infected person—even someone who has no symptoms—or by consuming food or water that has been touched by someone who failed to wash after using the bathroom or changing a diaper. Food-safety experts do not consider it compulsive for parents or anyone caring for small children to wash their hands eight to ten times a day. If you are not changing diapers, you probably ought to be washing your hands five or six times daily, more if necessary. Whenever possible, use liquid soap from a pump bottle or wall-mounted dispenser. Bacteria have been shown to grow on bar soap.

Here is a checklist of when to wash your hands:

- Before and after eating
- After touching raw meat, poultry, or fish
- After grocery shopping
- After handling garbage
- After using the toilet
- After changing a tampon or sanitary pad
- Before and after sex
- Before putting in contact lenses
- After touching an animal or changing a cat's litter box
- After touching blood or body fluids
- After gardening or working outdoors (even if gloves were worn)
- After every diaper change (the child's hands should be washed, as well)
- After blowing your nose or covering your mouth while coughing or sneezing
- Before and after treating cuts or touching someone who is sick or hurt
- After using public transportation or spending time in a crowded public space

A study published in the May 1997 issue of *Family Medicine* concluded that school-age children who followed a schedule of washing their hands at least four times a day experienced fewer sick days due to gastrointestinal (GI) symptoms (nausea, vomiting, diarrhea, abdominal cramps) compared with a control group that followed no particular hand-washing schedule. While it is impossible to know whether those GI symptoms had been food-borne, gastrointestinal distress is a hallmark of most food-borne infections.

Hand drying. Thorough hand drying is probably as important as hand washing to prevent food-borne illness. Researchers at Auckland Hospital in New Zealand have demonstrated that rinsed but undried hands can transfer tens of thousands of bacterial cells to food. Their follow-up studies, reported in the December 1997 issue of *Epidemiology and Infection,* found that drying hands for ten seconds using a clean cloth towel followed by air drying for twenty seconds achieved a 99.8 percent reduction in the amount of bacteria "translocated" (moved from one place to another) to skin, and a 94 percent reduction in bacteria translocated to food.

A fresh paper towel is more hygienic than a cloth towel that is damp or has not been recently laundered. Hot air dryers are efficient in theory, but people rarely use dryers long enough. The same New Zealand researchers showed that ten seconds of drying with a cloth towel removed 96 percent of the water from a person's hands, but it took forty-five seconds to achieve this level of dryness using an air dryer. Often, people complete the hot air-drying process by wiping their hands on their clothes, or they apply makeup and comb their hair while their hands are still moist, studies have shown. All these activities can potentially spread bacteria.

Hand-washing guidelines. By following the hand-washing guidelines below, you can dislodge virtually all microorganisms from your hands:

1) Turn on the faucet and allow the water to run until it is warm. (Cold water works, but warm water will help the soap cut grease faster.)
2) Wet hands.
3) Apply soap to hands.
4) Rub hands together vigorously for ten to twenty seconds. Be sure to wash both sides of your hands, including knuckles, between fingers, and under fingernails, where bacteria like to hide, especially if your hands are grimy. Loosen or take off rings so you can wash underneath.
5) Rinse your hands completely under running water.
6) Turn off the faucet with a paper towel.
7) Dry your hands completely with a paper towel, hot air dryer, or freshly laundered towel.
8) Open the bathroom door with a paper towel (or toilet tissue if no paper towels are available), or use your back or shoulder to push it open, if possible. You can recontaminate your hands if people who touched the doorknob before you did not wash their hands.

Some *Shigella* strains may survive on surfaces such as countertops and faucets. A December 1997 study in the *Journal of Clinical Microbiology* reported a *Shigella sonnei* outbreak involving a group of laboratory technologists. The technologists stopped getting sick after the lab was cleaned with a disinfectant and a handle-free faucet was installed in the sink.

Antibacterial soaps. Soap and water don't actually kill microorganisms; they create a slippery surface so that the bugs slide off. Antibacterial soaps are overkill and probably don't affect viruses. Some scientists are concerned that antibacterial soaps may spur the rise of more resistant strains of bacteria. Plain soap and warm water are all you need, even if you are caring for an infant or a sick person.

When there is no soap and water available, an alternative is using an antimicrobial, waterless hand sanitizer, such as Purell Brand Instant Hand Sanitizer and Bath & Body Works Instant Antibacterial Hand Gel. Sold in many supermarkets, drugstores, and department stores, these products are made principally of ethyl alcohol mixed with emollients and other ingredients. These products are not effective in removing dirt and grime; they are formulated to kill germs. Purell has been used in hospitals and restaurants as a hand-washing supplement for ten years, says company spokeswoman Megan Pace. According to results of test-tube studies provided by Gojo Industries Incorporated of Cuyahoga Falls, Ohio, the makers of Purell, the product reduces the amount of *E. coli* and *Streptococcus pyogenes* by 99.9 percent. Waterless sanitizers should be allowed to evaporate completely from the skin before handling food. One manufacturer's literature suggests using waterless sanitizer before and after eating in a car, after using a public phone, after riding public transportation, or after changing a diaper where soap and water are not readily available. At least one brand of waterless sanitizer is scented to appeal to children. Daniel Y. C. Fung, Ph.D., food science professor at Kansas State University, recommends that people who eat out carry a small bottle of alcohol-based hand sanitizer and use it to clean hands before and after eating, after using the bathroom, and after shaking someone's hand before a meal.

Washing little hands. An article in *Children's Healthwatch* published by the Mayo Clinic suggests that parents and other caregivers supervise hand washing by young children. The article also suggests ways to make hand washing fun by posting a chart that children can mark each time they wash, setting hand-washing goals, and rewarding children for meeting those goals for several days in a row, or by singing songs about hand washing.

While she was a professor at Purdue University/Calumet campus studying hand-washing habits in a daycare center, Joann Niffenegger wanted to graphically demonstrate the power of soap and water to children. So she coated their little hands with petroleum jelly then sprinkled on some ground nutmeg. When the children tried to wash away the nutmeg "germs" with mere cold water, it didn't work. Only warm water and lots of soap and scrubbing did the job. "Just telling kids (to wash hands) isn't enough," Niffenegger asserts. "It's an abstract notion."

Annie Condit says she is working hard to impart the soap-and-water routine to her four children. "I'm always after my children to wash their hands and not touch the doorknob after washing," she says. "When they're washing, I try to make my children aware of what their hands touch and how dirty the water is after they wash. I also try to make them aware of how much better their hands look and feel when they're clean."

Grocery Shopping

Safe grocery shopping is a partnership between you and the store's manager and staff. No matter how carefully a food item has been handled before it arrives at the store, mistakes by you or store employees can render it spoiled or contaminated. For example, if you notice a leaking package of ground beef or outdated chicken and report it, you may help prevent your family or another family from becoming ill. If the conveyor belt at a checkout lane is visibly wet or dirty, don't be shy about asking the cashier to wipe it down with some disinfectant spray and a paper towel. An old piece of lettuce or other rotting food wedged in the bottom of a shopping cart should be pointed out to the manager on duty. Because harmful microorganisms may dwell on grocery-cart handles, consider beginning your shopping trip by wiping down your cart's handle with a premoistened towelette, especially if you are at high risk for infectious diseases; and wash your hands after grocery shopping.

The condition of a supermarket's bathroom can speak volumes about an establishment's attention to sanitation. Is there enough toilet paper, soap, and paper towels? Do the toilet and hot-water tap work properly? Is the floor littered with cigarette butts? If the bathroom appears clean, there is a better chance that the stockroom, deli, and meat-packaging and salad-bar preparation areas are also clean. If the bathroom is filthy or poorly equipped, remember that the same bathroom is probably being used by the employee slicing the turkey breast you will feed your children for lunch tomorrow. If, after complaining to the manager, the bathroom is still dirty on your next visit, report the situation to your local health department—or consider finding somewhere else to shop.

There are other ways to keep your food safe while grocery shopping, according to Rutgers Cooperative Extension, from which many of the following suggestions originate:

- Do not purchase frozen foods that are stored above the frost line in a freezer case.
- Select well-sealed packages of chicken or meat from the bottom of the case, where the temperature should be coolest.
- Always keep meats, poultry, and seafood separate from other items. *Salmonella, E. coli,* and other disease-producing microorganisms thrive in raw meat juices. If these juices leak in your shopping cart or grocery bag, other food items can become contaminated. To prevent this from happening, place each meat and poultry item in

its own plastic bag, and close the bag tightly with a knot or twist tie. Do the same with frozen packages of meat and poultry. Many grocery stores put produce bags in the meat and poultry sections for this purpose.

- Don't buy fresh meat or poultry packages that are dripping or wet to the touch. If you get blood or juices on your hands, wash up before touching any other grocery items, particularly produce and other ready-to-eat foods.
- Don't allow small children to touch packages of raw meat, fish, or poultry. If they do, wash their hands right away.
- If you can smell your way to the seafood section, keep on walking. Fresh fish shouldn't smell "fishy."
- Cooked shrimp or other ready-to-eat seafood should not be in the same case as raw seafood. Cooked seafood should not even touch the same mound of ice that is holding raw seafood.
- Food-service workers who handle seafood or precooked foods such as deli items should wash hands or change gloves between each customer. They should also wash hands or change gloves between handling raw and ready-to-eat products. Wearing plastic gloves is no guarantee of safety; in fact, gloves may convey a false sense of security. Washing hands fastidiously is better than wearing gloves. Although processed deli items are sanitary if properly refrigerated, carelessness can introduce contaminants. For example,

contamination can occur if deli items are laid on a bare counter or scale.

- Don't buy deli items unless the meat slicers, countertops, cutting boards, scales, and floors appear clean. Deli workers should be wearing reasonably clean clothes and a hat or hair net.
- Buy locally grown produce, or produce grown in the United States, whenever possible. Ask the produce manager where a particular fruit or vegetable was grown if you are unsure. At this writing Congress was considering legislation requiring all produce sold in the United States to be labeled as to where it was grown.
- Pick up produce, dairy products, eggs, and frozen food toward the end of your shopping trip, after you have gotten your shampoo, napkins, coffee, canned goods, and other nonperishables.
- Select dairy products (with the latest "sell-by" date you can find) from the back of the refrigerated section, where it is colder.
- Retrieve raw meat, poultry, and seafood just before checking out. This is particularly important during the warmer months. As mentioned earlier in this book, bacteria can multiply to dangerous levels in two hours if the meat is not refrigerated. This window of safety shrinks to one hour in April through October in most parts of the country, and year-round in Florida, southern California, and other Sun Belt states.
- If you grocery shop on a hot day, consider

putting a well-insulated cooler with ice packs in your car (not in the trunk where temperatures are highest) to hold refrigerator and freezer items until you can get them home. To avoid fumbling through your bags in the parking lot, have the cashier bag all the cold items together. Bagging cold items together throughout the year provides a measure of insulation, regardless of whether you are using a cooler.

- Place all fresh fruit and vegetables in plastic bags and seal them with a knot or twist tie. This is particularly important if they are to be eaten uncooked.
- Buy packaged, precooked foods only if the package is completely undamaged—no tears, holes, or open corners. Buy ready-to-eat refrigerated foods only if you find them in a refrigerated case and if they feel cool to the touch.
- Purchase products labeled "keep refrigerated" only if they are displayed in a refrigerated case.
- Read freshness dates on cheese, milk, meats, and other perishable goods, and buy items marked with the latest date you can find. Don't buy any items if the "sell-by," "use-by," or "pull-by" date is expired.
- Buy frozen products only if they seem to be frozen solid.
- Never buy dented, bulging, rusted, or leaking canned goods. If you discover a damaged can at home, return it to the store. In rare instances, these cans harbor harmful bacteria.

- Don't buy more refrigerator items than you have room to store in your refrigerator.
- Don't run errands on your way home if you have a car full of groceries. Remember that even short stops during hot weather may let your groceries warm up to unsafe temperatures, leading to spoilage and the possibility of food poisoning.
- If you prefer to tote your groceries home in cloth or string bags, launder the bags periodically.

As you heighten your awareness of food safety, safe shopping practices will become second nature.

Safe Food Storage

The next time you open your refrigerator, take a brief survey of expiration dates. You'll probably find at least one product that has outlasted its usefulness. Even if you haven't yet opened the package, the safest move is to throw it away. Certain food-borne pathogens, such as *Vibrio* species (found primarily in mollusks but can also contaminate any seafood), and *Listeria monocytogenes* (found on raw vegetables, soft cheeses, fish, poultry, and meats) can multiply at refrigerator temperatures. (For guidance on how long to keep these foods, see the safe storage list on pages 227–29.) Molds can also thrive in a refrigerated environment. If the expiration date is illegible, or you are in doubt about the freshness of any food, discard it, especially if it looks or smells "off." (Remember—it is unwise to taste food as a method of determining quality or safety. Once you put it in your mouth, it may be too late.) Cutting mold off salami, fruits, or vegetables and eating the remainder is generally ill advised. Many molds produce toxins within the food that cannot be seen or smelled. A few dollars wasted is a small price to pay to prevent illness.

Of course, refrigerating food is useless—and dangerous from both a food-safety and quality standpoint—if the temperature inside your refrigerator exceeds 40°F. "The best thing a consumer can do is take the time to have their refrigerator temperature evaluated, and to run that refrigerator as cold as they can without freezing the food," advises David K. Bandler, professor of food science at Cornell University. He suggests buying an appliance thermometer and dipping it in ice water. If the ice water brings the reading to 32°F, then the thermometer is properly calibrated. Leave it on a refrigerator shelf at all times. Running your refrigerator between 36° and 38°F, Bandler says, "will give you more protection than anything else you can do." Freezers should be maintained at 0°F, or below.

Here are some more food-storage guidelines:

- Place cold items in the refrigerator or freezer as soon as you return from the grocery store. Meat and poultry should be kept in plastic bags or covered with plastic wrap or aluminum foil before being stored in the refrigerator or freezer. Putting the package on a plate in the refrigerator will prevent any juices from dripping onto other foods.

- Allow some space between packages stored in a refrigerator. This lets cold air circulate more freely to cool food as quickly as possible.
- Do not store highly perishable foods (including eggs, milk, fresh meat, fish, and poultry) on the refrigerator door. Door temperatures can be slightly higher than inside the appliance's main compartment.
- Avoid placing aluminum foil directly over acidic foods such as fruit, juice, or tomato sauce. The acid can "eat through" the aluminum, exposing the food to bacteria in the air.
- If juices from raw meat or poultry touch a table or countertop, the area should be disinfected as soon as possible to prevent other food or food containers from being contaminated. Use paper towels dampened with a mild chlorine bleach solution (one to two tablespoons of bleach per gallon of water) or commercial kitchen cleanser. "Cleaning with chlorine in dilute solution is probably the best thing to prevent contamination with food-borne pathogens," says Barry Swanson, Ph.D., professor of food science at Washington State University.
- Wipe down refrigerator shelves and drawers with a bleach solution or commercial cleanser once a week. If juices from raw meat, fish, or poultry drip onto the shelf, clean the area right away with your bleach solution and paper towels. Unless they are going to be cooked, food items that come in contact with the spill should be tossed. Any food containers that come in contact

with the spill should be washed with soap and warm water.

- Read all food labels to find out which items need continuous refrigeration and which need refrigeration only after opening.
- Store chopped garlic-and-oil mixtures in the refrigerator. "I don't recommend homemade garlic-in-oil, only the store-bought kind, which is acidified when it is made, to control botulism," cautions Rutgers' Donald W. Schaffner, Ph.D., a nationally known food-safety expert.
- Never reuse a plastic bag, wrap, or aluminum foil that covered raw meat, poultry, or fish.
- Discard any frozen food item that completely thawed en route from the store.
- Be sure to refrigerate custard-type pies, including homemade pumpkin pie. In the wintertime, you can cover pies and place them outside or in a garage for a few hours, as long as the ambient temperature is as cold as the refrigerator—40 degrees or below.
- Freezing food can extend its shelf life, but freezing unsafe food won't destroy bacteria that have already grown. Therefore, freeze only fresh or freshly cooked food as soon as possible; don't let it sit around your kitchen for hours before storing it. Freezer wrap, freezer bags, and aluminum foil will help preserve the quality of frozen food.
- Foods that develop "freezer burn" (dehydration from being exposed to air in the freezer) are safe

to eat, even if they have lost some color, flavor, or texture, according to the July 1996 *Tufts University Diet and Nutrition Letter.* To prevent freezer burn, cover food with freezer-safe cling wrap. For extra protection, add a second layer of freezer wrap or aluminum foil.

- Write the date on foods that you freeze. "Processed meats like bacon and hot dogs will keep only for a month or so, and small cuts and poultry parts will keep on the order of three to six months," the *Tufts Letter* states. "But big cuts of meat, such as roasts and whole chickens, will remain top quality for up to a year if wrapped properly. Foods like bread and cheesecake, on the other hand, will keep only for two to three months."
- If you lose electricity, or your refrigerator stops working, don't open the door. If the problem continues for more than a couple of hours, you should cook any raw meat or poultry right away. Frozen food should be transferred to a working freezer after two hours. If you lose power for more than two hours, and your refrigerator thermometer reads over 40 degrees, throw away your perishables.

The *Tufts Diet and Nutrition Letter* points out that a freezer works most efficiently when it is full because the foods have an insulating effect on one another. The *Tufts Letter* suggests placing some crumpled newspaper or paper bags in an almost-empty freezer. "This trick also works when a power outage threatens your frozen foods. Quickly pack the freezer with paper, and the food inside should stay frozen for twenty-four hours."

Here is a list of safe storage periods for selected refrigerated foods:

EGGS	
Eggs, fresh in shell	3 weeks
Eggs, in shell, hard-boiled	1 week
MAYONNAISE	
Store-bought, opened	2 months
RAW HAMBURGER, GROUND AND STEW MEAT	
Hamburger and stew meat	1 to 2 days
Ground turkey, veal, pork	1 to 2 days
HAM	
Ham, canned, opened, labeled "Keep Refrigerated"	3 to 5 days
Ham, fully cooked, whole	7 days
Ham, fully cooked, half	3 to 5 days
Ham, fully cooked, slices	3 to 4 days
HOT DOGS AND LUNCH MEATS	
Hot dogs, opened package	1 week
Lunch meats, opened package	3 to 5 days
BACON AND SAUSAGE	
Bacon	7 days

Sausage, raw from pork, beef, chicken, or turkey	1 to 2 days
Smoked breakfast links, patties	7 days

Fresh Meat (Beef, Veal, Lamb, and Pork)

Steaks	3 to 5 days
Chops	3 to 5 days
Roasts	3 to 5 days
Variety meats (tongue, kidneys, liver, heart, chitterlings)	1 to 2 days

Meat Leftovers

Cooked meat and meat dishes	3 to 4 days
Gravy and meat broth	1 to 2 days

Fresh Poultry

Chicken or turkey, whole	1 to 2 days
Chicken or turkey, parts	1 to 2 days
Giblets	1 to 2 days

Cooked Poultry, Leftovers

Fried chicken	3 to 4 days
Cooked poultry dishes	3 to 4 days
Pieces, plain	3 to 4 days
Pieces covered with broth, gravy	1 to 2 days
Chicken nuggets, patties	1 to 2 days

SEAFOOD	
Raw	1 to 2 days
CHEESE	
Natural, processed	4 to 8 weeks
Cream, Neufchatel, ricotta, cottage	1 to 2 weeks
FRUIT	
Canned, opened	1 week
VEGETABLES	
Canned, opened	3 days

Safe Food Preparation and Cooking

From a bacterium's standpoint, your kitchen has all the "right stuff"—time, temperature, moisture, air, and nourishment—for a comfortable lifestyle. The ambient temperature is a balmy 70 degrees, or perhaps as high as 100 degrees when your stove and oven are on in the summer. Kitchen towels, aprons, sponges, dishcloths, pot-scrubbers, and the like can sit around damp for hours at a stretch. Used to wipe hands, utensils, and countertops, these materials usually contain food residue, fat, and protein—a virtual feast for pathogens. So, for bacteria, which can reproduce about every twenty minutes, it is party time. Fortunately, you have the power to end the party.

One rule of thumb is to change into a clean, dry apron daily and to use only freshly laundered kitchen towels, disinfected sponges, and paper towels. Washing your hands after each step in the food-preparation process can prevent cross-contamination. Also remember to wash again after interruptions, especially if you have to use the toilet. You may be immune to the bacteria from your own intestinal tract, but you can easily transfer these bugs to food that will be eaten by someone who is not immune, explains Bessie Berry, manager of the USDA's Meat and Poultry Hotline. If you have an open cut or sore on your hand or wrist, you can protect both your food and yourself by wearing plastic gloves or covering the wound with a bandage during meal preparation. When you are ill, do not prepare food that is to be eaten by others. Always avoid sneezing or coughing into food.

Unwashed raw fruits and vegetables can also spread pathogens around your kitchen, especially if the produce was exposed to animal feces, including bird droppings, in the field or orchard. Scrubbing with a brush under running water will remove most of the dirt, soil, debris, insects, manure, pesticides, and bacteria. You may opt to peel your fruits and vegetables, but you sacrifice the fiber, vitamins, and minerals that may be present in the outer layer. Most fruits and vegetables that can be peeled can withstand a good scrubbing to remove the majority of contaminants that may be present on the outside. Grapefruits, kiwis, bananas, and other fruits with inedible rinds should also be scrubbed before peeling or cutting. This proviso includes cantaloupes and other melons. If the rind is contaminated, a knife can transport contaminants from the outside to the edible interior. For the same reason, wash and dry strawberries well before removing the stems and slicing them.

Rinsing raw meat, chicken, or seafood, on the other hand, is not recommended. Rinsing cannot remove all the bacteria—but it can spread it around your sink. Your sink then becomes a vehicle for cross-contamination when it comes time, for example, to wash your salad greens.

Cutting boards. Cutting boards can also easily become vehicles for cross-contamination. Bacteria can contaminate sandwiches or salad ingredients that are placed on a cutting board that was not washed after being used to prepare raw animal products or unwashed produce. The big debate has been over which kind of cutting board is best, wooden or plastic. For years the FDA recommended plastic boards based on the observation that they were easier to clean. Research has shown, however, that wooden boards are less likely to spread contamination. Also, bacteria can hide in knife scars on plastic boards and defy washing with soap and water.

In experiments carried out in the early 1990s, Dean O. Cliver, Ph.D., a food microbiologist at the University of California at Davis, and colleagues purposely contaminated seven different types of wooden cutting boards and four types of plastic boards with *Salmonella, Listeria,* and *E. coli.* After three minutes, almost 100 percent of the bacteria on the wooden boards had died or could not be recovered, but almost none had died on the plastic boards. Overnight, the bacteria on the plastic boards multiplied, yet no bacterial samples could be recovered from the wooden boards. Maple or similar closed-grain hardwood cutting boards have microscopic pores into which the bacteria are quickly pulled, leaving none on the surface, the researchers said. According to food safety expert Donald W. Schaffner, research from other laboratories has shown that bacteria absorbed into wooden cutting boards may be in a "viable but nonculturable" (VNC) state. VNC bacteria are alive but cannot be grown on normal microbial test media. But if they somehow manage to get into your intestines, Schaffner says, they can still make you sick. "We still use plastic boards in my house and run them through the dishwasher," he says. "We use wooden boards only for cutting bread."

Scientists may disagree on which cutting board is best, but from the consumer's perspective the bottom line is this: It is the board's cleanliness and condition—not its composition—that matters most. Cutting boards should be sanitized as soon as possible after each use to avoid spreading germs. This can be done in a number of ways:

Wooden or Plastic Cutting Board

- Run the board through the dishwasher (using a "sani" cycle if you have one); or
- Scrub the board with hot, soapy water, then pour a kettle of boiling water over it. (Hand-scrubbing alone may not remove all the germs from a knife-scarred plastic board); or

- Spray the board with vinegar, then hydrogen peroxide. (See pages 157–58.)

Wooden Cutting Board

- Microwave on high for ten minutes. (Of course, this method is practical only if your board is small enough to fit inside your microwave.) Microwaving does not work on plastic boards because the plastic won't get hot enough to kill bacteria.

Plastic Cutting Board

- Immerse the board in a bleach solution for a minute or so, then rinse well with running water. If you choose this method, you should probably add a little more bleach than the standard one tablespoon per gallon of water, in order to kill all the bacteria on the cutting board. Bleach, even at full strength, will not sanitize wooden cutting boards because the organic composition of the wood neutralizes the disinfectant quality of the bleach, according to Patrick J. Bird, dean of the College of Health and Human Performance at the University of Florida, who recently wrote about cutting boards for a column in the *St. Petersburg Times*. Bird adds that cutting boards should be kept dry when not in use because any lingering bacteria may grow in a moist environment.

All cutting boards, whether they are made of wood, plastic, or marble; impregnated with pesticides; or labeled "antibacterial," should be carefully washed after each use to prevent cross-contamination. When your plastic board becomes badly scarred from knife slashes, it makes good sense to toss it and buy a new one. Cutting boards are designed to be tools—not family heirlooms.

Temperature safety. Another important rule is to keep food out of the temperature "danger zone"—that 40°F to 140°F window in which bacteria can proliferate rapidly on many foods. *Therefore never defrost meat or poultry at room temperature.*

- The safest way to defrost is in the refrigerator or microwave oven. If you defrost in cold water, make sure there are no tears in the plastic wrap (rewrap if necessary), then fully immerse the item in cold water. Change the water every thirty minutes until the item is defrosted.
- Food should be cooked immediately after defrosting; don't place defrosted food in the refrigerator to be cooked later.
- Prechill salad ingredients, canned tuna and salmon, hard-boiled eggs, and cooked pasta before preparing and serving. Any bacteria that find their way into these dishes will have a harder time multiplying if the food is cold.
- For the same reason, whole watermelons and cantaloupes should be placed in the refrigerator

or on ice before being used. Melons should be consumed within four hours of being cut open.

In general, you should prepare meals as close as possible to serving time. If you prepare foods hours or days in advance, you give pathogens more opportunity to multiply in your food.

Meat Thermometer. The best way to ensure that meat is cooked to a safe temperature is by using a meat thermometer—an essential kitchen utensil. In June 1997, the USDA Food Safety Inspection Service (FSIS) issued a consumer advisory to use a thermometer when cooking hamburger. Traditionally, hamburger doneness was evaluated by sight; if the interior of the patty was no longer red, the burger was presumed safe to eat. New research indicates, however, that color is an unreliable indicator of doneness. In some cases, ground meat oxidizes and turns prematurely brown before a safe internal temperature of 160°F has been reached, the FSIS points out. In other cases, ground meat patties safely cooked to 160°F or above may remain pink in color, says Thomas J. Billy, FSIS administrator. Using a meat thermometer is the most reliable way to reduce the risk of food-borne illness in ground beef, he adds. The FSIS offers these additional thermometer tips:

- Use an "instant-read" thermometer, which is designed to be used toward the end of the cooking time and to register a temperature in about fifteen seconds.
- To check thermometer calibration, place the

stem into a cup of boiling water. It should read 212°F. Most thermometers have a calibration nut under the dial that can be adjusted.

- Check the stem of the instant-read thermometer for an indentation showing how deeply it must penetrate the meat to get an accurate temperature reading. Most digital thermometers will read the temperature in a small area of the tip. By contrast, dial types must penetrate about two inches into the food.
- The thermometer should penetrate the thickest part of the hamburger. If the patty is thin, the thermometer may be inserted sideways.
- To measure the temperature of whole chicken the thermometer should be inserted into the innermost portion of the thigh.

Never use a mercury-in-glass thermometer to measure the temperature of food. Mercury is a highly toxic metal, and if the glass cracks or breaks, the mercury can leak into the food.

If you place a meat thermometer in a burger and learn that it is not yet cooked, your thermometer may be contaminated when you remove it from the patty. So be sure to wash the thermometer's tip with soapy water or a mild bleach solution before testing the patty's temperature again.

In addition to the 160-degree minimum internal temperature for hamburger, meat loaf, and other ground-beef dishes, the USDA recommends these minimum internal temperature food-safety guidelines (all temperatures are Fahrenheit):

Fresh beef, veal, and lamb (roasts, steaks, and chops): medium rare: 145°; medium: 160°; well done: 170°

Fresh pork (all cuts including ground pork): medium: 160°; well done: 170°

Poultry: ground chicken/turkey: 165°; whole chicken/turkey: 180°; whole bird with stuffing: 180°; stuffing: 165°; breasts, roasts: 170°. Thighs and wings should be cooked until the juices run clear.

Ham: fresh raw: 160°; cooked (to reheat): 140°

Egg dishes (i.e., quiche): 160°

It takes about one second for bacteria and other pathogens to be killed at the aforementioned temperatures.

Poultry Cooking Tips. Placing poultry on a wire rack in the oven allows heat to surround the bird and cook it more evenly. To help heat reach into joints of the bird, do not truss (bind) whole poultry legs; wings should be folded akimbo. A stuffed precooked bird should be refrigerated as soon as you get it home; it should be reheated and eaten within two hours of purchase. According to the FSIS, in its 1996 publication *Turkey Basics: Stuffing,* if you stuff a raw turkey yourself, mix the stuffing just before it goes into the turkey, and stuff loosely—about 3/4 cup of stuffing per pound of turkey. The FSIS recommends that the stuffing be moist, not dry, since heat destroys bacteria more rapidly in a moist environment. Never store a stuffed raw bird in the refrigerator; place it in a preheated oven (at least 325°F) immediately after it is stuffed. After it is cooked, check the stuffing temperature to make sure it has reached 165°F. Also check the temperature in at least three different places on the bird, even if the turkey has a pop-up temperature indicator. Your goal is to kill *all* the bacteria, not most of them.

Toward that end, never partially cook or brown foods to cook later because any bacteria present would not have been destroyed, the FSIS says. When using a Crock-Pot, cut meat or poultry into small chunks and stir periodically to ensure complete cooking.

In an advisory issued in December 1997, the FSIS warns consumers to be especially careful about cooking poultry and ham products that have been injected with a basting, curing, or flavoring solution that is used to enhance the food's color, flavor, or shelf life. The FSIS points out that bacteria from the surface may be carried into the product's interior as the solution is injected in the processing plant—creating an environment that could foster bacterial growth. If the label indicates the presence of a basting, curing, or flavoring solution, it is even more critical to cook the food to the proper internal temperature, the FSIS says.

Sausage. When cooking sausage, boil it for several minutes, cut it in half lengthwise, then grill or fry it until a safe internal temperature (160° F) is reached.

Seafood. Fin fish should be cooked until it flakes easily with a fork. Crustaceans such as shrimp and lobster should be cooked until they turn red. Bivalve shellfish should be cooked until their shells open. If possible, cook and eat seafood within twenty-four hours of purchase. Eating seafood raw is one of the biggest culinary risks you can take. According to published data, the marine bacterium *Vibrio parahaemolyticus* accounts for 45 to 60 percent of all food-related infections in Japan, with disease rates peaking in August and September. From 1980 to 1993, Japan tallied a total of 4,063 outbreaks involving 116,838 cases of food-borne illness (five of them fatal). In 1994, there were 224 outbreaks involving 5,849 cases, and in 1995, 245 outbreaks involving 5,515 cases.

Spice it up. Have you ever wondered why countries near the equator have a more aromatic cuisine than those in northern latitudes? It is not only because chili peppers and cumin are endemic to southern regions, or that spicy foods make the palate feel more alive. A pair of Cornell University researchers believes there are evolutionary explanations as well: to protect themselves against microbial attack, spice plants contain natural antibiotics that are also capable of killing or inhibiting many food-borne bacteria, including disease-causing *Salmonella enteritidis, Campylobacter jejuni, E. coli,* and *Clostridium botulinum*.

According to one of the researchers, Paul W. Sherman, Ph.D., of Cornell's Department of Neurobiology and Behavior, people living in hot climates came to use spices over thousands of years, either consciously or unconsciously, to prevent food-borne infections and to retard food spoilage. He and colleague Jennifer Billing reported their findings in a forty-six-page paper published in the March 1998 *Quarterly Review of Biology*.

To investigate why spice use differs among cultures and countries, and why spices are even used at all, Sherman and Billing quanitified the use of forty-three spices in the meat-based cuisine of thirty-six countries representing every continent and stretching from the Netherlands to South Africa. They analyzed a total of 4,578 recipes from ninety-three cookbooks, compiled data on the antibacterial properties of each of the forty-three spices, and looked at temperature and rainfall in each country. Their analysis revealed that the spices used most abundantly in recipes from the hottest regions inhibited 75 to 100 percent of bacteria species against which they were tested. Garlic, onion, allspice, and oregano are among the spices that inhibited 100 percent of bacteria species; thyme, cinnamon, tarragon, cumin, cloves, lemon grass, bay leaf, peppers, rosemary, and marjoram inhibited 75 percent or more. Even comparatively mild spices such as parsley, basil, and nutmeg inhibited about 50 percent of food-borne bacteria.

While the proximate reason spices are used is to enhance food palatability, the researchers concluded that the ultimate or evolutionary reason is most likely that spices help cleanse foods of pathogens and thereby contribute to the health, longevity, and reproductive success of people who find their flavors enjoyable. Prior to the widespread availability of refrigeration and artificial chemical additives for preservation, using spices in recipes lowered the risk for food-borne infections—no matter where people lived.

"I think that much of what we do with foods, including salting, drying, cooking, and now spicing, has to do with avoiding food-borne illnesses and food-borne pathogens," Sherman says. "These pathogens are all competing for the same resource—either dead plant or animal matter—as we are." Pathogens, he says, are trying to consume the food source as fast as they can, and they're putting out, in many cases, toxins that presumably evolved as a mechanism to compete against other food "consumers," such as bacteria, fungi, insects, and vertebrates. Some of these toxins are metabolic by-products, but others are manufactured in a way that will cause the meat or vegetable to quickly putrify, thereby ensuring that only the toxin-producing bacteria will be able to consume it. In many cases, these toxins can cause illness in human hosts.

Eating and enjoying spicy foods is so ingrained among tropical and subtropical populations that they may not be aware of the health benefits of their culinary habits. That realization isn't even necessary, Sherman points out, drawing an analogy to sex. "Sex is enjoyable, and we don't necessarily think about it only in terms of having babies. We evolved this behavioral liking of sex, presumably because those people who enjoyed sex left behind more progeny than those who didn't enjoy sex. So any tendency to like something that's good for you will be passed to the next generation, whether or not there's a cognitive knowledge of it. On the same idea, I'm suggesting that people come to like the taste of spices because their parents taught them to like it, and it was good for them.

A second possible explanation involves "learned taste aversions." In his paper, Sherman notes that when people eat something that makes them ill, they tend to subsequently avoid that taste. Adding a spice to a food that caused nausea might alter its taste enough to make it palatable again; it might also kill the microorganisms that cause the illness, thus rendering the food safe for consumption. "By this process," Sherman writes, "food aversions would more often be associated with unspiced (and unsafe) foods, and food likings would be associated with spicy foods, especially in places where foods spoil rapidly."

Microwave safety. Microwave ovens do not heat food evenly and may therefore leave "cold spots" that can allow bacteria to survive and cause food-borne illness. To prevent this problem, the FSIS offers the following tips from its 1997 publication, *Microwave Food Safety:*

- Remove food from store wrap prior to microwave defrosting. Foam trays and plastic wraps are not heat stable at high temperatures. They may melt or warp, possibly causing chemicals to migrate into food.
- Don't microwave food in cold storage containers, such as margarine tubs, whipped topping bowls, cheese containers, because they can warp or melt from hot food and possibly cause chemical migration.
- Arrange food items uniformly in a covered dish and add a small amount of liquid. Under a cover, such as a lid or vented plastic wrap, steam helps destroy bacteria and promote more uniform heating.
- When microwaving meat, measure the meat's temperature in at least three different places to

evaluate whether it is thoroughly cooked. If any readings are too low, continue to microwave until a safe temperature is reached throughout.

- Debone large pieces of meat before cooking in a microwave. Bone can shield the meat around it from thorough cooking.
- Rotate or stir the food once or twice during the cooking period to facilitate even heat distribution. For the same reason, turn large food items upside down during microwaving.
- Cook large pieces of meat on medium power (50 percent) for longer times. This allows heat to conduct deeper into meat without overcooking outer areas.

Avoid recontamination. Even when meat or poultry is cooked to the proper temperature, it can easily be recontaminated by hands, plates, or utensils such as spatulas and tongs, and cutting boards that were not properly washed and dried after coming in contact with the raw product. Never serve cooked meat on a platter that wasn't washed after holding the raw product. Raw patties should be cooked simultaneously. Placing a raw hamburger patty in a frying pan next to a patty that is done can contaminate the cooked burger.

Don't use a fork to turn chicken, steaks, or other solid meats as they are cooking. While contamination can exist throughout ground beef, bacteria are generally confined to the surface of solid cuts of meat (see Chapter 4). This is why medium-rare steaks and chops, which are brown on the outside and pink on the inside, rarely cause food-borne illness. However, by piercing the meat with a fork, you can transfer surface contamination to the sterile interior—making that medium-rare steak more dangerous to eat.

If your recipe calls for marinating, do so in a covered dish inside the refrigerator. Don't marinate longer than the food's recommended storage time (see list on pages 227–29). Marinades make foods more flavorful, but they do not destroy bacteria. Throw out leftover marinade; it is almost assuredly contaminated with bacteria. Don't taste any meat, poultry, seafood, or egg dish while it is raw or cooking.

Egg alert. Don't give in to the temptation to taste cake batter and cookie dough that contain fresh eggs; store-bought cookie dough is safe to eat raw because its ingredients are pasteurized. If a recipe calls for an egg to be eaten raw (i.e., Caesar salad, hollandaise sauce), a safer alternative is pasteurized egg substitute. Homemade eggnog is also safer when pasteurized eggs are substituted for fresh. Store-bought eggnog is pasteurized. Make sure French toast and scrambled eggs are cooked throughout, especially if they are to be served to children, the elderly, or anyone with a weakened immune system. Eggs are safest when hard boiled, hard scrambled, or cooked over hard. Runny eggs have not reached a sufficient temperature to ensure that bacteria have been destroyed.

Bag lunch safety. Bag lunches should be kept at 40°F or cooler. This means using an insulated bag with an ice pack or storing it in a refrigerator until lunchtime. You can make your own ice pack by filling a margarine tub almost to the top with water, putting on the cover, and freezing it. Or you can put some ice cubes in a small zipper-lock plastic bag. Instruct your children to throw out any uneaten food at school. Don't leave a bag lunch in a warm car or in direct sunlight, even if it contains a cold pack.

Safety on the grill. Be extra cautious about food eaten at picnics and barbecues. For one thing, temperature control is more difficult outdoors. Keep all perishable food wrapped in plastic and in a cooler en route to the picnic. If loose ice is being used, be sure the cooler has proper drainage so the melted ice cannot spread contamination.

If there will be no bathroom facilities nearby, bring along some wet wipes or waterless sanitizer to clean your hands before and after preparing food at the picnic site. Use a clean grill, or clean off major debris and let the grill and coals heat up thoroughly before cooking (coals should be gray to red). Do not use the same spatula, fork, or tongs that touched raw food to remove cooked food from the grill unless you wash the utensils with hot, soapy water. Avoid putting a cooked item onto the same plate that was used for raw items.

All cooking temperatures should be measured with a good meat thermometer. Whole muscle meats such as steak, rib roast, and kabobs should be cooked to at least 145°F; hamburgers to at least 160°F (internal temperature), and chicken to 170°F. Be sure to insert the thermometer into the thickest part of the meat. If you don't have a meat thermometer, cook the burgers until they are obviously well done and the juices are running clear;

the same goes for barbecued chicken. Some cooks microwave, boil, or partially bake chicken before barbecuing so the inside will cook before the outside burns. If you decide to try this two-step method, be sure to put the chicken on the grill immediately after step one. Regardless of which cooking method you used, always examine the interior of chicken before you bite; if you see any hint of pink meat or slightly bloody juices, put the chicken back onto the grill or into the microwave. If you marinate chicken or ribs in barbecue sauce, never brush leftover sauce onto food that is cooking or is about to be eaten.

Another danger of eating outdoors is insects and vermin. Flies, roaches, rodents, and other creatures commonly carry bacteria—including *Salmonella* and *E. coli* picked up from fecal matter, unclean water, or sewage—on their bodies. Insects and animals can spread these bacteria to any surface their body parts touch, including your utensils, plates, towels, clothing, apron, countertops, tables, or other surfaces. To reduce the risk of this happening, don't set up your picnic near weeds and high grasses, or where garbage isn't removed regularly. If possible, cover your picnic food with a netted tent, wax paper, or aluminum foil until it is eaten. Burning a citronella candle nearby can help ward off certain insects.

Don't take picnic leftovers home; leftovers should be thrown in garbage cans at the picnic site. As noted earlier, in the summer bacteria can grow in raw meat or poultry after one hour, if the food is not refrigerated.

If you are throwing a party, use a warming table or hot plate to keep hors d'oeuvres at a safe temperature, and keep cold foods such as half-and-half or cream on ice. Hard cheese can safely stay at room temperature for several hours.

Safe handling of leftovers. Store leftovers in the refrigerator or freezer as soon as possible after your meal. If the leftovers' temperature rises above 40°F for two hours or more, bacteria can grow even if the food was prepared and cooked safely. Refrigerating or freezing does not destroy bacteria that has grown in leftovers. Only reheating leftovers or precooked meals to 165°F or higher will do that. To promote rapid cooling of warm leftovers in the refrigerator, place the food in shallow containers.

Cleaning up. Antibacterial cleansers, dish soap, and sponges are all the rage. Yet there is no proof that using these products will reduce your chances of getting sick. Antibacterial sponges can still spread germs to hands and food-contact surfaces, unless they are cleaned after each use. According to the *Wellness Letter,* plain soap or detergent is just as effective in the kitchen as an antibacterial product. In fact, some scientists suspect that overuse or misuse of certain disinfecting products may spur the evolution of stubborn strains of bacteria. In one recent study inspired by a high school science project, a strain of *E. coli* bacteria was fed a diet of pine oil, a main ingredient in many kitchen and bathroom cleansers. After exposure to the pine oil, the *E. coli* tolerated a two- to eight-times higher dose of common antibiotics, including tetracycline and ampicillin, compared with *E. coli* that had not been exposed to pine oil. Other types of cleansers could possibly have the same effect, says Stuart Levy, a specialist in bacterial resistance at Tufts University School of Medicine, and one of the study's coauthors.

The study, which appeared in the December 1997 issue of the scientific journal *Antimicrobial Agents and Chemotherapy,* found that the pine oil somehow switched on multiple antibiotic-resistant genes called *mar. Mar* apparently protect the bacteria from attack by pushing the antibiotics and disinfectant out of the *E. coli* cells. When the scientists deactivated one of the *mar* genes, the pine-oil resistance was turned off.

Despite his findings, Levy indicates that he is not concerned about the emergence of dangerous antibiotic-resistant *E. coli* strains as a result of using household disinfectants. He is, however, concerned about antibacterial properties being added to hand lotions, plastics, toys, and other products. Antibacterial products are not needed, he says, except perhaps when caring for a sick person. "Let's reserve them for this use and avoid propagation of resistant strains," he says.

Daryl Minch, family and consumer sciences educator with Rutgers Cooperative Extension, stresses the importance of using disinfectant cleaners as directed on the label. Usually you have to wipe the surface clean first, with or without the product. Once the surface is visibly clean, you then wet the surface again with the product, and you must leave it wet, usually for around ten minutes. If it dries sooner, there's no guarantee there's been enough contact to kill any bacteria. Or, if you spray the counter and wipe it and think, "I've done a good job"—that's not going to do it either. It has to remain wet for a while.

A mild chlorine bleach solution is a very effective and inexpensive alternative to commercial cleaning fluids, Minch points out. But you shouldn't mix a bleach solution on Sunday and expect it to remain potent all week. You have to make up the solution fresh each day because otherwise the bleach breaks down and becomes ineffective, she says. To make a solution capable of killing household germs, you need only one tablespoon of bleach per gallon of water (or $1^1/_2$ teaspoons per half-gallon) for nonporous surfaces such as Formica, porcelain, steel, acrylic, and tile. Dilute bleach solution is still quite strong, Minch says. "There's no need to go hog wild and assume if a little is good, a lot is better."

Sponges and dishcloths can spread illness-causing bacteria such as *Salmonella* to every other surface they come in contact with: countertops, cutting boards, and dishes, states the April 1997 *Tufts University Health and Nutrition Letter*. Sponges may not completely dry between uses, and the moisture helps bacteria multiply. Putting sponges in the dishwasher may make matters worse if all the bacteria in the sponge are not killed and water drips from the sponge onto clean dishes, the *Tufts Letter* continues. Sponges with antibacterial properties are safer than regular sponges, but do not offer a complete guarantee of safety.

After each use, sponges should be cleaned with liquid dish soap and running water, or soaked in a mild bleach solution and rinsed, or doused with boiling water. If you prefer to launder kitchen sponges, be sure to use bleach and hot water. (Do the same for dishrags, dishcloths, aprons, or any clothing that gets stained or splattered with food.)

A fifth way to sanitize a sponge is to microwave it on high for one minute. "Sometimes my sponges start to smell after only one week of use," says Janet C., a mother of three. "Then I heard about microwaving them. After only one minute in the microwave, they don't smell anymore. It's amazing." To eliminate all risk of cross-contaminating your kitchen with a sponge, use paper towels instead.

Even if your sink and sponges are disinfected, bacteria-laden particles of food can cling to your drain basket and even inside your drain. The drain basket can be sanitized in the dishwasher, or you can sanitize it along with the drainpipe with a mild bleach solution or boiling water.

Handle Reptiles with Care

According to Jonathan Mermin, M.D., and Fred Angulo, D.V.M., Ph.D., medical epidemiologists at the CDC, all reptiles and amphibians probably carry *Salmonella,* which may not make the animal sick but can cause human disease and even death. Multiple studies have shown that reptiles and amphibians in zoological parks and in the wild carry *Salmonella,* and research further suggests that reptiles in pet stores also carry the bug.

"What this means from a public health point of view is not that zoos and aquariums should not involve these animals in their educational programs, but that a few precautions should be taken," Mermin says. Children under age five, pregnant women, and people with weakened immune systems should avoid touching reptiles and amphibians. Children age five and older should wash their hands with warm, soapy water—with adult supervision, if needed—after touching reptiles or any animals, for that matter.

Children with pet snakes, iguanas, turtles, frogs, or other reptiles or amphibians should be warned never to put their hands in their mouths while handling these creatures. Pet reptiles should be kept away from food-preparation areas. Reptile cages, accessories, and food dishes should not be washed in the kitchen sink. Day-care centers and other facilities occupied by young children should not keep reptiles or amphibians.

Like reptiles, live chicks, ducklings, and goslings have been shown to shed *Salmonella* in their stools. Health officials typically issue a warning every Easter advising adults to avoid giving children these birds as pets.

It may not require direct contact with a reptile to contract disease. According to a study by the CDC published in the May 1998 *Journal of Pediatrics,* up to sixty-five people, mostly children, developed salmonellosis after touching a wooden barrier surrounding a Komodo dragon exhibit at a metropolitan zoo in 1996. A dragon infected with *Salmonella enteritidis* "stood in fecally contaminated mulch and frequently placed [its] front paws on the tops of the barriers," the researchers noted in their study. "The [zoo] visitors frequently touched these same barriers and most likely became infected when they later placed their contaminated hands in their mouths or cross-contaminated something they were eating." People who had washed their hands after touching the barrier were far less likely to get salmonellosis, the researchers said.

Person-to-Person Infection

Touching the skin of an infected person and then putting your hands on your food or in your mouth can transmit germs, including ones that are usually foodborne. According to the Lois Joy Galler Foundation for Hemolytic Uremic Syndrome, a Long Island-based support and fund-raising group, *E. coli* O157:H7 is an example of a bacterium that can be "touch-borne," meaning a child can contract HUS by simply touching someone who ate one of these contaminated foods—and didn't wash their hands after a resultant bout of diarrhea or gastroenteritis. In general, hand-to-hand and hand-to-mouth contact is required; merely bumping into an infected person will not cause bacteria to "jump" onto you.

Zap Trap

When a common housefly gets zapped by an electronic trap, millions of bacteria on the surface of the insect are scattered into the air because the zapper causes the flying pest to explode. The bacteria are flung six feet or even farther if a wind current is present, meaning that the zappers potentially spread more disease than they prevent. This bit of disconcerting news was presented by James Urban, professor of biology at Kansas State University, during a recent meeting of the American Society for Microbiology. As noted by Reuters news service, which reported Urban's comments, bacteria covering the legs and bodies of houseflies come from human or animal wastes. Urban advises against using an electronic zapper. When eating in snack bars, backyards, or other places where these devices are used, keep yourself and your food at least six feet away from the zapper. Urban also recommends that bug zappers be situated so they are not exposed to air currents from fans, vents, or open windows.

10

DINING OUT SAFELY

On January 1, 1992, state health officials in New Jersey enacted a controversial rule: restaurants in the Garden State could no longer serve raw or undercooked eggs. No more eggs over easy. No more fresh eggs in the Caesar salad dressing or hollandaise sauce. Omelets and French toast had to be cooked to a minimum internal temperature of 140° F. Violators would be fined between $25 and $100. Health officials cited the rising number of Americans contracting *Salmonella* from contaminated food, mainly raw or undercooked eggs, as a reason for the ban.

A number of other states, including Connecticut, Maryland, Massachusetts, the Carolinas, and Virginia, had a similar rule. But New Jersey's egg ban became a national joke. Ridiculed by politicians, the ban inspired copy editors to write such headlines as, "Hearing Held on Breaking Up State's Runny Egg Rule" and "The Great Egg Scramble." On the *Tonight Show,* host Johnny Carson quipped that you could buy an Uzi in New Jersey, but there was a ten-day waiting period to get a Caesar salad. (Actually, Uzis were illegal in New Jersey.) In his state-of-the-state address, then-Governor Jim Florio characterized the egg ban as an example of silly government intrusion into people's lives. The Associated Press quoted a Florio spokesman as saying that the governor "did not intend to make light of anything." Nonetheless, Florio invited the media to watch him eat an order of sunny-side ups at a diner in Bordentown. By mid-February, the health department had reluctantly agreed to relax the rule and let New Jersey restaurant customers order eggs any way they pleased.

New Jersey's runny-egg fiasco exemplified what Tom Montville, Ph.D., professor of microbiology and chairman of the Department of Food Science at Rutgers University, calls "the fundamental dilemma in food safety. From a public health standpoint, the runny egg law was a good public health measure. But people felt it infringed upon their freedom." Consumers want to be protected against food-borne illnesses, but they do not want anybody telling them what they can and cannot eat.

As New Jersey health officials observed, the percentage of reported food-borne diseases contracted outside the home has increased significantly in recent decades. Around 1970, about 60 percent of food-borne-disease outbreaks occurred outside the home—primarily in restaurants and institutional settings. By the 1990s, that figure had jumped to 70 or 80 percent, estimates Mitchell Cohen, M.D., director of the Division of Bacterial and Mycotic Diseases at the Centers for Disease Control and Prevention. One reason is that people are twice as likely to report a food-borne illness to their local health department if they believe it was contracted at a restaurant as opposed to in their own kitchens. This may skew the data and give the impression that eating out may be more risky than it really is.

Societal changes explain why Americans now spend 40 percent of their food dollars in restaurants, up from 27 percent in 1970. With the rise in dual-income and single-parent households, there are simply fewer stay-at-home moms cooking dinner every night. Rushing to send their children off to school in the morning, some working parents find it easier to grab a bagel and coffee at a convenience store instead of preparing their own breakfast. Millions of employees eat their midday meal in a restaurant or cafeteria. At the end of the day, they may be too tired to cook. Two-income families can probably afford to eat out more often, anyway.

This explosive growth in the food-service industry has placed unprecedented pressure on local health departments, which may lack adequate funding and personnel to inspect restaurants as frequently and intensively as they would like. A recent survey of forty-five states, counties, and cities around the country by the Center for Science in the Public Interest (CSPI), a nonprofit consumer group, found that only half of those agencies surveyed inspected restaurants at least twice a year. And none of the agencies enforced all twelve "key" FDA food-safety guidelines in restaurants, according to the CSPI report entitled, "Dine at Your

Own Risk: The Failure of Local Agencies to Adopt and Enforce National Food Safety Standards for Restaurants." Only 13 percent of the agencies surveyed enforced the FDA Food Code's recommended cooking temperatures for pork, eggs, fish, and poultry; and only 64 percent of the agencies required that hamburgers be cooked to a high enough temperature for a long enough time period to destroy *E. coli* O157:H7. The survey's results led CSPI Director of Food Safety Caroline Smith DeWaal to conclude that "weak, poorly enforced state and local regulations result in millions of cases of food poisoning—and several thousand deaths—each year." The survey has been attacked by the National Restaurant Association for being flawed and unscientific. To judge for yourself, you may obtain a copy of "Dine at Your Own Risk" by sending $10 to CSPI Dine Report, Suite 300, 1875 Connecticut Avenue, Washington, DC 20009-5728.

Heather Klinkhamer, program director for Safe Tables Our Priority (STOP), which was founded in 1993 by families and friends of people who were sickened or who died from ingesting *E. coli* O157:H7 in meat, says that some members of her nonprofit group claim that local health departments "get a lot of political pressure to acquiesce to local businesses—to overlook violations, or to give them a second chance."

As your awareness of food-safety issues increases, it is easy to feel vulnerable in a restaurant. To some extent, these feelings of vulnerability are valid. After all, you are eating food prepared by total strangers behind closed doors. You have no idea whether the restaurant employees are always washing their hands after using the bathroom. You do not know whether the chef preparing your food has hepatitis A, or if your server received adequate food-safety training.

"We're at their mercy," says Klinkhamer.

On the other hand, restaurants have strong moral and financial incentives to handle food properly. One Florida restaurant owner was hit with lawsuits and received death threats after state health officials confirmed 198 cases of salmonellosis and 873 complaints of illnesses of varying degrees from customers who ate at the West Palm Beach restaurant during two days in August 1995. Health officials closed the restaurant for ten days while the owner overhauled food-handling procedures, according to an account in *Restaurants and Institutions,* a trade magazine. "My bottom line went to zero or minus" during the incident, the owner disclosed to colleagues during a board meeting of the National Restaurant Association.

The good news is that diligence on the part of the consumer can sometimes overcome food-safety lapses that may occur in an eating establishment. One key is selecting menu items wisely. This chapter will show you how. Additionally, you will learn how to evaluate food-handling practices in daycare facilities and nursing homes. About 20 percent of *E. coli* O157:H7 outbreaks reported in the United States occur in these two settings. According to STOP, in 1995 alone, at least sixty-five cases of *E. coli* infection were transmitted in daycare centers.

This chapter will also help you reduce your risk for traveler's diarrhea and suggest ways to ensure that you are eating safely while at work.

Restaurants

Despite your lack of control over how a restaurant stores and prepares your food, you can look for clues—some subtle, some not so subtle—that show how much the establishment heeds the rules of food safety. There are also several steps you can take and questions you can ask to make your dining experience as safe as possible.

Perhaps the best indication of a restaurant's safety record is its most recent inspection report. A restaurant ought to be willing to display this document regardless of whether they must do so by law. In Georgia, restaurants are required to post near the door or somewhere near the cash register—where it is readily visible to the patron—what their sanitation scores were for the last three inspections, says Michael Doyle, Ph.D., director of the Center for Food Safety and Quality Enhancement at the University of Georgia. "There's no uniformity among states in terms of how they deal with sanitation inspections. There needs to be, and that's being addressed," Doyle says. "But I think it's the best evidence one can go by" when deciding whether to eat in a particular restaurant.

In Maryland, the restaurant industry recently teamed up with the state health department to launch a voluntary food-safety certification program, and several other states have indicated a desire to follow suit. Participating restaurants, daycare centers, retirement homes, health-care facilities, and other food-serving establishments enroll existing employees and new hires in a two-hour food-safety course once a year. If 75 percent or more of the employees complete the course, the establishment receives a "Seal of Commitment" that can be displayed to customers and used in advertising, like the Good Housekeeping Seal of Approval. Maryland's initiative is designed in part to balance negative publicity surrounding food-poisoning scares and to reassure the public that the industry is working to improve food-safety practices among professional food handlers, according to Lisl Wilkinson, executive director of the Maryland Hospitality Education Foundation, which helped develop the program. Within four months of the announcement of Maryland's initiative, twenty eating establishments had enrolled, and eight to ten other states—including Tennessee, Oklahoma, New Jersey, Texas, Delaware, Pennsylvania, and Florida—had requested information on Maryland's program, Wilkinson notes.

Food-safety experts are hard-pressed to advise consumers on whether to eat in a restaurant that has been a source of a previous outbreak of food-borne illness. In some cases, restaurants become more careful about food safety, compared to those that have never gotten into trouble. In other cases, very little changes. Klinkhamer says there is no guarantee that the public will find out about an outbreak caused by restaurant food in the first place.

In general, a restaurant's reputation or expense has no relationship to the safety of its food, experts say. Many fast-food chains like McDonald's do an excellent job with food safety, says Donald W. Schaffner, Ph.D., extension specialist in the Department of Food Science at Rutgers University. "It's that dimly lit gourmet restaurant I'm most worried about," he adds with a grin.

Food scientist Tom Montville notes that there are "very clean, cheap places, and there are not-so-clean expensive places."

So, before sitting down to order, Montville makes a beeline for the bathroom. A clean bathroom, he believes, can be a "pretty good indication" that the kitchen is clean. Cohen of the CDC agrees. "I don't have any data, but I would think that if you have a poorly maintained bathroom, it would be a general reflection of other aspects of cleanliness. So that's probably a reasonable indicator."

Signs of neglect include overflowing garbage pails, filthy floors, stopped-up drains, lack of hot water, and dirty toilet seats. If there is no soap in the bathroom, Montville advises against eating in that restaurant because it suggests employees are not washing their hands with soap. In some places, employees use the same bathroom as customers; in other places they do not. You just never know. One time when Montville pointed out the absence of bathroom soap to the restaurant manager, the response was a dismissive, "Not a problem," to which Montville replied, "No, it's a problem." Needless to say, Montville and his party ate elsewhere that night.

If the restaurant passes the bathroom test, look to see if it passes the dining-room test. Are busboys wiping tables and counters with dirty cloths or sponges, or are they using disinfectant spray and clean paper towels? Unless dutifully disinfected, dishcloths and sponges can become vehicles for bacterial contamination.

If you can see food being prepared, as in a pizza parlor or sushi bar, notice whether the food handlers are taking precautions. At sushi bars, says Kansas State University food-safety expert Daniel Fung, the customer should touch the glass display panel. "If it is very cold, the sashimi (raw fish) are safe. But if the glass is not cold, do not eat raw seafood there." Evidence of safe food handling was clearly apparent during one recent evening at the Golden Teapot, a restaurant in Hamilton, New Jersey, that specializes in Japanese and Chinese cuisine. The staff was shorthanded that night, so in addition to preparing sushi, the chef had to take telephone orders and operate the cash register. After helping a customer pay for a takeout order, the chef washed his hands with soap and water, put on a fresh pair of plastic gloves, then washed his gloved hands with soap and water *again* before touching any food.

Another thing to look for in a restaurant dining room is how the servers are dressed. Their clothing should be reasonably clean and their hair netted or pulled back. Hands should be washed between clearing and setting a table. Certain disease-causing microorganisms, including *Staphylococcus aureus* and hepatitis A virus, can be transmitted via tableware. Before eating, Fung recommends dipping your spoon, fork, or chopsticks in hot tea or soup to sanitize the utensils.

Ordering from the menu is an important, critical control point for consumers. Caitlin Storhaug, spokesperson for the Washington, D.C.–based National Restaurant Association, a trade group representing 175,000 restaurants nationwide, advises people to consider their current health status when determining what to order for themselves or their children. If you feel you are in a high-risk group because of your age or a medical condition, ask your doctor, nurse, or nutritionist if there are items, such as raw oysters, that you should avoid. For example, people with liver impairment, especially alcohol-induced liver disease, can become gravely ill if exposed to *Vibrio vulnificus,* a bacterium found in some raw Gulf Coast oysters and occasionally in raw clams. As mentioned in previous chapters, pregnant women, anyone whose immune system is weak or immature, the elderly, and people who are ill or who are taking antibiotics or antacids are especially vulnerable to food-borne infections.

Even if you are not in a high-risk group, you can still get sick from eating unsafe food. "I get calls from people who were very healthy but were hospitalized with a food-borne illness," STOP's Klinkhamer says. "Just because you're a healthy adult doesn't mean you're not at risk. It's just that your risk is reduced."

When ordering, do not be afraid to ask questions. If you are in the mood for Caesar salad, you have every right to ask the server or manager if the dressing is made with raw eggs or pasteurized eggs. (If bottled dressing is used, it is probably safe.) "If you like vinegar-based dressings, pour a lot of it on your salad. That will kill some bacteria," advises Fung. He also believes there is some truth to the killing power of hot sauces such as Tabasco or other brands. "I carry a small bottle with me when I travel and always sprinkle it generously onto my food, especially sunny-side up eggs."

The safest approach is to order only foods that are least likely to cause a problem, advises Cohen of the CDC. Even if the chopped meat you order is contaminated when it arrives at the restaurant or is cross-contaminated by the chef, the bacteria will be killed if it is served fully cooked (to 160°F) and hot.

Montville says he is generally wary of any dish that requires extensive handling, such as chicken-salad and turkey-salad sandwiches, which are frequently made from leftovers. These types of dishes are very microbiologically sensitive; bacteria love to grow in them. "Because of its nutrient sources and the way it's all mixed together, something like a chicken-salad sandwich is a much better growth environment for bacteria than a well-done hamburger," he says. "If you were to buy a well-done hamburger and a chicken-salad sandwich and leave them in your car on a summer's day, I would eat the well-done hamburger, but I wouldn't touch the chicken-salad sandwich."

Another likely reservoir for bacteria are roasted ducks that are hung at room temperature in some traditional Chinese restaurants and delicatessens. If you order duck in one of these restaurants, it should be hot when served.

In fast-food outlets, Montville says, it is wise to special order. A burger ordered without the regular complement of condiments usually must be cooked on the spot, which diminishes the possibility of getting a burger that's been sitting under a warmer for several hours. In sit-down restaurants, well done is the safest way to order hamburger, meat loaf, Salisbury steak, or other chopped-meat dishes. Steaks are usually safe cooked medium-rare as long as the outside is seared. Montville also specifies how he wants his fish cooked. "I've had people in very good restaurants tell me, 'We normally prepare fish medium rare.' I say, 'Not mine.' I always order fish at least medium well done." If cooked shellfish shells are open, any bacteria have probably been destroyed.

Do not be embarrassed or intimidated over sending a dish back to the kitchen if the food is not hot enough, not cold enough, or not cooked well enough. In general, any time you see steam rising from your food or drink, you can consider it safe for consumption, says Fung. "Cold dishes should be really cold in your mouth." Salad ingredients, for example, should be chilled and served on cold plates, he says. Be suspicious when the hot food is warm or the cold food is cool.

If a poultry dish is the least bit bloody or pink, it should be sent back to the kitchen. Eggs are also safest when they are well cooked. Quiche should be firm, not runny.

"Sometimes you have to stand up for your rights in a restaurant," says Christine Bruhn, Ph.D., who studies consumer attitudes and perceptions about food safety. "If you see someone sticking their fingers inside your glass or cup when they're serving it, or doing other gross things, be brave; tell them: 'That's gross, give me another one please,' or better yet, tell their boss."

Street vendors. Unlike restaurant chefs, people who sell hot dogs, gyros, and other foods on the street do not work behind closed doors. Their food-handling practices are right out in the open, and safety-savvy consumers will know which vendors to avoid. When buying meat, chicken, or egg products from a food cart, make sure the food is steaming hot and thoroughly cooked before you eat it. Avoid vendors who touch raw or semicooked items with their bare hands. Also look to see if the vendor is routinely washing his or her hands and utensils with soap and water (a sink should be available).

In New York City, which has more food carts than anywhere else in the country, roughly 40 percent of vendors handle foods that can be hazardous, according to a recent story in *The New York Times*. Tests arranged by *The Times* on a random sampling of chicken, burgers, and kebabs from vendors' carts showed significant undercooking in thirty-nine of fifty-one cases—meaning that bacteria would not be eliminated. As the story points out, the combination of risky fare, poorly equipped facilities, and unsafe handling create a potentially dangerous situation for food-cart patrons.

Dr. Robert Gravani, a food scientist at Cornell University who tested street fare at the request of *The Times,* ordered food, watched its preparation, paid for it, and immediately tested it with a digital thermometer. Of eight hamburgers tested, just one was fully cooked. Of eleven chicken dishes, only two met health standards. Of five beef kebabs tested, none measured up, the newspaper reported in the May 17, 1998, article. Gravani told *The Times* that he did not once see vendors wash their hands or utensils. Rather, many sinks were bone dry and inaccessible. In one case, Gravani ordered a well-done hamburger, then watched as the vendor picked up a half-thawed patty with his bare hands, tossed it on his grill, and used the same unwashed hands to open the bun it was served in. A nearby test showed the patty was still raw inside and 131 degrees instead of the 160 degrees needed to kill *E. coli,* were it present.

Gravani was quoted as saying, "I wouldn't let my children eat that."

Leftovers. If you are planning to catch a movie after dinner, do not even consider bringing home a doggy bag, unless you happen to have an ice chest in your car. In many instances, the food has already been at room temperature for at least an hour as you finished your meal. That leaves only another hour to safely get your leftovers into your refrigerator or freezer. Before eating, leftovers should be thoroughly reheated.

Daycare Centers

Around noon on July 21, 1993, eighty-two children in two jointly owned Virginia daycare centers put down their puzzles and toys and trotted over to the dining area for lunch. On the menu was chicken fried rice, peas, and apple rings. About two hours later, twelve children and two daycare staff members had developed the typical symptoms of food poisoning—vomiting, abdominal cramps, and diarrhea.

An investigation by the Lord Fairfax Health District traced the illnesses to *Bacillus cereus,* an infectious and toxin-producing bacterium, in the chicken fried rice. According to investigators, the rice had been prepared the day before at a local restaurant, where it was left to cool at room temperature before being refrigerated. The next morning, the rice was pan-fried in oil with pieces of cooked chicken and delivered to the daycare centers at around ten-thirty in the morning. At the centers, the rice was held without refrigeration until being served at noon—without being reheated.

Fortunately, the outbreak was short-lived; all the victims recovered about four hours after the onset of symptoms. Had the infectious agent been *Salmonella* or *E. coli* O157:H7, the outcome could have been tragic.

Working parents look for many different things in a daycare facility: a clean, bright, stimulating atmosphere conducive to learning and socialization; a trained, caring staff; safe playground equipment; a convenient location; reasonable rates; and flexible hours. With all these factors to consider, it can be easy to overlook how a daycare center handles food. Because young children are considered a high-risk population for food-borne infections, food safety must be a top priority in daycare centers. Pat Kendall, Ph.D., of the department of Food Safety and Human Nutrition at Colorado State University at Fort Collins, suggests asking some of the following questions when shopping for a daycare center, or if you are concerned about food-handling practices in a facility you are already using:

- *If I bring in formula or breast milk from home, how do you handle the bottles?* Except when they are being used, bottles of formula or milk should be in a refrigerator. If you notice bottles sitting on tables, desks, or countertops, it could be a tip-off that the center is probably not going to handle your child's food safely, either.
- *Where are the babies diapered?* One of the biggest areas of concern is fecal contamination from one baby to another through the diapering process or

by transferring microscopic bits of fecal material to food or to eating utensils. Daycare facilities should have a designated diapering area that is separated from the kitchen and eating areas. Dining tables and countertops must never be used to change diapers. Try to spend some time watching diapers being changed. Are staff members washing their hands with soap and water before and after each change? Does someone change a diaper then handle cookies or crackers or prepare lunch without washing their hands in between? Is the diapering area sanitized after each change?

- *How is the kitchen organized?* The kitchen in a daycare facility should be clean and orderly. Take pause if you see cups and dishes piled in the sink, or milk or food left out between meals. There should be adequate refrigerator space to accommodate all the bottles and food a facility needs.
- *How often are children's hands washed?* Babies should have their hands washed after each diaper change, and before and after eating. Daycare staff should supervise toddlers and preschoolers as they wash hands after using the toilet, and before and after eating, and after going outside to play.
- *What kind of universal precautions do you follow?* Be very concerned if the daycare provider has no idea what you mean by "universal precautions," which refers to a set of procedures used to prevent the transmission of infectious diseases. Among other things, daycare workers should

wear latex gloves when treating a cut or scrape, and they should safely dispose of blood-exposed bandages and gauze. *Staphylococcus aureus* and other infectious microorganisms can be transmitted from an open wound to foods such as bananas or peanut-butter-and-jelly sandwiches if universal precautions are not followed.

- *How are sick children prevented from spreading their illness to others?* Sick children should be separated from the healthy population throughout the day—while sleeping, playing, and eating. A sick child could spread illness by sneezing into or touching another child's food.
- *Are parents encouraged to prepare meals or snacks at home to be shared with other children in the daycare?* Food is a great way to encourage camaraderie and teach children about sharing, but it is also an easy way to spread food-borne infections. For example, a mother could have hepatitis A, not realize it, and spread the infection to dozens of children in a daycare setting through a batch of homemade cookies. A food-borne illness outbreak in the Fort Collins, Colorado, public school district led officials there to ban homemade food from being served to groups of students, according to Kendall.

Nursing Homes

Nursing homes are ripe territory for food-borne diseases. These long-term care facilities house a frail, elderly population living in close quarters. Residents may have fecal incontinence, weakened immune function, or stomach linings that are no longer acidic enough to kill pathogenic bacteria. Some are further weakened by chronic diseases. As a result, outbreaks in nursing homes are not unusual.

One of the most common food-borne bacteria found in nursing homes is *Salmonella.* In one recent salmonellosis outbreak in a Maryland nursing home, 50 of 141 residents (35 percent) became ill, 9 were hospitalized, and 4 died as a result of the infection. Because their immune systems tend to be weak, the frail elderly are at high risk of becoming seriously ill or even dying after exposure to food-borne pathogens. Nursing homes, therefore, need to follow food-safety guidelines very closely to protect their residents.

For clues to a nursing home's food-safety practices, spend some time in the kitchen and dining hall. If possible, observe an entire meal, from preparation to soup to nuts to cleanup. The kitchen should look clean and the staff well organized. The dining area should not have a foul odor. Food should not be placed on tables an hour before residents sit down to eat. Food ought to be hot and fresh when it is served. If a food's color is off or it looks as though it has been reheated for several hours, it may not be safe to eat, Kendall says. Tables should not be wiped with sponges or dishcloths unless a disinfectant is being used.

Ask about a facility's dining hours. "Sometimes they'll bring the residents into the dining room and let them sit there half a day before taking them back to their rooms," Kendall says. "What that says to you is that the staff is overworked and may take shortcuts in terms of food safety."

Also take note of situations in which residents too sick or weak to come to the dining hall eat meals in their rooms. Is the hot food hot and cold food cold when it is delivered? How long does the food sit in the room before it is consumed? If it is midafternoon and the lunch tray has not been picked up yet, it could indicate that the nursing-home staff is not very organized. And just because a dish is covered does not mean the food is free of harmful bacteria.

On the Job

A major meeting is taking place in the conference room at work. All the bigwigs from your company's corporate headquarters are there. A catered lunch arrives at 11 A.M., and the meeting drags on for hours. When the huddle is over, the leftovers—including shrimp cocktail, cheese, a variety of salads, fresh vegetables and dip, cold cuts, and rolls—are offered to you and your staff. Everything looks delicious. But should you eat it? Probably not, advises Alice Henneman, M.S., R.D., an educator at the University of Nebraska Cooperative Extension and editor of *FoodTalk,* an on-line newsletter.

Perishable food such as dairy products, eggs, meat, poultry, and seafood should not be eaten if they remain at room temperature for more than two hours. The same rule applies to cut fruits and vegetables. In the November 1997 edition of *FoodTalk,* Henneman discusses these other food-safety problems that typically occur in the workplace:

- **Double dipping.** People can spread germs by dipping a chip, taking a bite, then dipping the chip again. Beat the double-dippers to the dip and put enough dip on your plate to enjoy with all your chips.
- **Dirty dishcloths.** Encourage the use of disposable paper towels to wipe off the sink and tables in the break room. Place your food on a napkin or paper towel rather than directly on the table surface. Coffee cups should be washed in hot, soapy water using a freshly cleaned dishcloth, then rinsed with hot water and allowed to air dry. Do not recontaminate clean dishes by drying them with dirty towels, especially towels that are also used to dry hands. If you have little control over how cups are cleaned, bring your own cup that you can clean appropriately.
- **Refrigerator madness.** If you put something in the refrigerator at work, eat it or toss it within a week. Post a sign to encourage co-workers to do the same.
- **Helping hands.** If there is no sign in the bathroom reminding people to wash up, make one and post it at eye level in bathroom stalls and in front of urinals. One idea adapted from a

Purdue University Cooperative Extension Service Conference is to draw an outline of your hands with your fingers spread apart. Let the caption read: "The 10 Most Common Causes of Food-borne Illness."

Traveler's Diarrhea

Gary L., a thirty-seven-year-old computer software consultant from Seattle, tried to be careful about what he ate during a three-week jaunt to Mexico in December 1997. He drank only bottled or filtered water, avoided salads and ice, and peeled his fruit before eating it. Then, ten days into his vacation, he purchased an innocent-looking yogurt, ice, and fruit drink called a liquado from a "not-so-sanitary-looking stand" in Puerto Angél in the state of Oaxaca. Gary, an intrepid traveler who backpacked across Eastern Europe for seven months in 1994, had enjoyed liquados from other places in Mexico and had done fine. That was not to be the case here. Twenty-four hours after enjoying his beverage, he became "a little nauseated—then 'blamm-o.' " He felt as if an avalanche was tumbling out of his gastrointestinal tract.

His stools, watery and explosive, came every few minutes to every few hours. He vomited almost as frequently and developed a fever that lasted for days. "I was starving and very nauseous at the same time," he recalls. "I remember thinking to myself, 'So this is how it feels to be dying. Isn't the Ebola virus like this?' I was also concerned about traveling because I had plans to leave in a few days, and I was in no position to ride a bus. It was so frustrating being on vacation and unable to do things."

Rarely life threatening, traveler's diarrhea (sometimes called bacillary dysentery, or dysentery) is reported by 20 to 50 percent of American tourists, according to the U.S. Centers for Disease Control and Prevention. Mexico is one of the high-risk destinations, along with most of the other developing countries of Latin America, Africa, the Middle East, and Asia, according to the CDC publication, *Health Information for International Travel 1996–97*. Southern European countries and a few Caribbean islands are considered intermediate risk for traveler's diarrhea (TD); the United States is a low-risk destination, as are Canada, northern Europe, Australia, and New Zealand.

Because Americans have a relatively safe and sanitary food supply, our resistance to certain microorganisms isn't as built up as that of people living in certain foreign countries, food-safety specialist Pat Kendall points out. Also, in developing countries, there are a lot of problems with sewage contaminating the drinking-water supply. So it is very easy, especially if you like to economize on food, to come in contact with germs that cause TD.

The CDC publication lists about a dozen species of bacteria that can cause TD; among them are enterotoxigenic *Escherichia coli* (ETEC), *Salmonella, Shigella, Campylobacter jejuni, Vibrio parahaemolyticus,* and *Yersinia enterocolitica.* Viral pathogens, specifically rotavirus and Norwalk virus, can also cause TD, as can the parasitic enteric (inflammation-producing) pathogens *Giardia lamblia, Entamoeba histolytica,* and others, although these account for a tiny percentage of cases. Despite all the known causes of TD, the culprit in up to 50 percent of cases cannot be identified. Some travelers whose stools test positive for a TD-causing pathogen never develop symptoms.

TD symptoms occur when a rapid, dramatic influx of unfamiliar microorganisms in the gastrointestinal tract overwhelms your natural defense mechanisms. The intestines become inflamed, producing diarrhea and vomiting until your immune system can churn out enough antibodies to fight off the infection. Gary's symptoms were uncommonly severe and of unusually long duration—about two weeks. More typically, TD results in four to five loose or watery stools per day for three to four days. Ten percent of cases persist longer than a week; about 15 percent include vomiting, and 2 to 10 percent may have diarrhea accompanied by fever, bloody stools, or both.

TD is slightly more common in young adults than in older people. The reasons for this difference are unclear, but may include a lack of acquired immunity, more adventurous travel styles, and different eating habits, the CDC publication states. As with all food-borne pathogens, people with weakened immune systems are particularly susceptible to TD. Prudent hand-washing practices are just as important when traveling as they are at home.

Prevention. The CDC recommends two approaches to preventing TD: strict dietary restrictions and limited prophylactic (preventive) use of Pepto-Bismol. Prophylactic antibiotic use has been shown to prevent TD in some situations but is not generally recommended because of the risks associated with these medications.

Meticulous attention to what you eat and drink when traveling abroad is the best way to prevent TD. As Gary found out, however, it can be extremely difficult to stick to recommended dietary restrictions especially when spending more than a few days in a high-risk area. A single drop of contaminated water on a piece of lettuce can harbor millions of TD-causing bacteria.

In general, the food and water served at resorts is safer than what you might find in local restaurants and in people's homes, says Kendall. Food that is thoroughly cooked and hot is unlikely to cause TD. Raw fruits and vegetables, water, and ice are far more likely to be contaminated. When traveling in high- and intermediate-risk areas, drink only bottled or canned beverages, and be sure to dry the bottle or can with a clean paper towel if it was chilled directly on ice. Safe beverages include bottled soda, beer, wine, hot coffee, hot tea, and water that has been boiled or appropriately treated with iodine or chlorine, according to the CDC. Only boiled, treated, or bottled water should be used for brushing teeth. You can buy water-purification tablets, such as Halazone, at most drugstores.

Peeling fruit or eating only canned fruit is another way to lower your TD risk, as is avoiding green salads and raw or undercooked seafood and eggs, especially if you are vacationing in a tropical region. Consuming unpasteurized fruit juice, raw milk, or any unpasteurized dairy product can raise your TD risk. If you are offered cheese, find out if it was made from pasteurized milk. If in doubt, do not eat it.

As alluded to above, the only nonprescription compound shown to be effective in preventing TD is bismuth subsalicylate—the active ingredient in Pepto-Bismol. Several controlled studies have found that taking two Pepto-Bismol tablets four times a day can decrease the incidence of TD by about 60 percent. There are, however, a few precautions: Such large doses of bismuth subsalicylate should not be taken for more than three weeks because the health effects are unknown. Side effects include temporary blackening of the tongue and stools, occasional nausea and constipation, and in rare instances, tinnitus (ringing in the ears), according to the CDC. Also, bismuth subsalicylate should not be taken by people who are allergic to aspirin or by anyone with kidney problems or gout. Since bismuth subsalicylate has aspirinlike properties, it should be avoided if you are taking aspirin for arthritis or another condition, or if you are taking anticoagulants or certain other medications. "Caution should be used in giving bismuth subsalicylate to adolescents and children with chicken pox or flu because of a potential risk of Reye's syndrome," the CDC publication warns. Bismuth subsalicylate is not recommended for children under the age of three. Check with your doctor if you are planning a trip to a high-risk region and wish to take bismuth subsalicylate to prevent TD.

Research suggests that prophylactic treatment with the antibiotics doxycycline, trimethoprim/sulfamethoxazole (TMP/SMX), trimethoprim alone, ciprofloxacin, or norfloxacin, is 52 to 95 percent effective in preventing bacterial TD in several areas of the developing world. However, the CDC does not recommend this approach because the risks associated with these antibiotics are potentially more serious than an episode of TD. Aside from allergy, antibiotic risks include rashes, photosensitivity (sunburn easily), blood disorders, and staining of the teeth in children. Additionally, antibiotics may induce the development of certain forms of colitis (inflammation of the colon), and yeast infections in women. Antibiotics provide no protection against viruses or parasites, and travelers who take antibiotics prophylactically may feel a false sense of security about the risks of consuming certain local foods and beverages, the CDC cautions. Travelers who wish to take prophylactic antibiotics should discuss with their physicians all the possible risks and benefits.

There are no vaccines available to prevent traveler's diarrhea caused by bacteria, although the FDA is weighing approval of a vaccine against rotavirus. Rotavirus causes a diarrhea and vomiting syndrome that kills an estimated three hundred infants a year in this country and up to a million infants worldwide.

Treatment. As mentioned earlier, TD usually subsides on its own within a few days. Until then, the biggest danger is dehydration, especially in children age two or younger. The CDC and World Health Organization recommend drinking an oral rehydration solution made of sodium chloride, potassium chloride, sugar, and trisodium citrate, or sodium bicarbonate mixed with boiled or treated water. Oral rehydration salts are widely available in most developing countries and in the United States. If you are traveling to a high TD-risk country with small children, it is probably a good idea to pack some Lytren, Pedialyte, Rehydralyte, or Resol in your suitcase. In some cases, these products are available in powdered form. After mixing the powder with boiled or treated water as directed, the solution should have a shelf life of twenty-four hours refrigerated and twelve hours at room temperature. Children with TD should take small, frequent sips of oral rehydration fluid and be fed a normal amount of starches, cereals, yogurt, fruits, and vegetables. Teenagers and adults do not necessarily need rehydration solution, but they should drink as much as possible and eat as little as possible for the duration of the illness.

Research studies have shown that bismuth subsalicylate (Pepto-Bismol)—one ounce or two tablets every thirty minutes for eight doses—reduces the rate of diarrhea. The same precautions mentioned for prophylactic use of bismuth subsalicylate should be followed. Taking Kaopectate or Imodium as directed may also provide symptomatic relief of diarrhea. Codeine or other natural opiates, diphenoxylate, and loperamide may provide temporary relief of abdominal cramps and diarrhea. These agents should not be taken, however, if there is fever, or blood in the stool.

But by suppressing diarrhea with drugs, you may make matters worse in the long run by allowing a bacterium such as *Salmonella* to linger in your intestine for a longer period of time. The one benefit of having diarrhea is that it ultimately flushes the bacteria out of your system. This is why many doctors prefer that TD be allowed to run its course without symptomatic treatment.

Directly attacking the infecting organism with antibiotics is often a better treatment option. According to the CDC, taking the correct antibiotic can often shorten a typical three- to five-day illness to twenty-four to thirty-six hours. The drugs most effective against TD are TMP/SMX, ciprofloxacin, norfloxacin, or ofloxacin taken twice a day for two or three days. Some of these antibiotics, including ciprofloxacin, are not currently approved for use in children.

Always consult a physician if your TD persists for more than four or five days, if there is blood or mucus in the stool, or if you develop a fever with shaking chills or dehydration.

Hepatitis A. Another widespread infection in the developing world is hepatitis A, which is spread by the fecal-oral route through close person-to-person contact or by ingesting contaminated food or water. Like TD, hepatitis A can cause nausea, vomiting, diarrhea, and fever. Other possible symptoms are jaundice (yellowing of the skin), muscle weakness, chills, respiratory symptoms, rash, and joint pain. Hepatitis A infection is relatively rare, affecting three to six out of every one thousand visitors to endemic areas such as Mexico, parts of the Caribbean, South and Central America, Africa, Asia (except Japan), the Mediterranean basin, Eastern Europe, and the Middle East.

The traditional antidote, immunoglobulin, is taken before or after exposure to the hepatitis A virus but before symptoms develop. A new vaccine called Havrix provides long-term immunity to hepatitis A in 80 to 98 percent of cases and is generally effective within fourteen days. To find out whether Havrix or any other vaccines are indicated for the country you are traveling to, call the CDC's International Travelers Hotline at (404) 639-2572.

11

CHALLENGES AHEAD:

The Future of Food Safety

By learning how disease-causing organisms can contaminate food and make you sick you are taking a significant step toward preventing food-borne infections. But educating yourself is only the starting point. The next vitally important steps are educating others about what you have learned and putting what you have learned into practice. Jill S., a registered dietitian who worked in the food-service industry for nine years, learned this the hard way when she got extremely ill in the spring of 1997. She believes it began at a buffet dinner in one of her favorite restaurants. Among the items Jill selected was cooked shrimp, which no one else in her family ate. She noticed that the shrimp on her plate were not chilled, but she consumed them anyway, along with some cocktail sauce.

Around 1:30 A.M.—about seven hours after leaving

the restaurant—Jill woke up with a start. She raced to the bathroom, and over the next couple of hours endured some fifteen episodes of painful, explosive diarrhea. By four in the morning, she was beginning to feel dehydrated; she collapsed every time she tried to stand up. Her husband, an internist, drove her to his office and attempted to restore her body's salt balance by administering saline solution intravenously.

"We left three sleeping children alone in the house; that's how desperate I was," Jill recalls, adding that she thought she was going to die. She recovered within a couple of days but still chides herself for eating the room-temperature shrimp: "I'm a nutritionist. I've studied food safety. I've worked in the field. I should have known better."

To this day, Jill does not know for sure that the shrimp she ate were contaminated. But that uncertainty has not stopped her from making some adjustments to her eating habits, and to those of her three children. For one thing, Jill's family will never patronize that particular restaurant again. Jill will no longer eat shrimp, or serve it to her family, unless she cooks it herself or buys a packaged shrimp cocktail off a pile of ice or from a refrigerated unit in the grocery store. She has not eaten from a buffet table since that terrible night and suspects she'll avoid buffets altogether for the rest of her life.

Almost a year after the shrimp incident, Jill was grocery shopping when her nine-year-old son, Jason, begged her to buy some buffalo wings from a steam table in the supermarket. The wings had been piled high in the center island of the buffet. "There was no way they were being kept above 140 degrees," Jill says. Jason didn't snack on buffalo wings that day.

People such as Robert L. Buchanan, Ph.D., the chief scientist for President Clinton's Food Safety Initiative, probably would be heartened by people like Jill. Her awareness of food-safety issues and her cautionary behaviors make Buchanan's Herculean task just a little bit easier.

Among the most pressing challenges for government, industry, and scientists, he says, is trying to anticipate the next food-safety crisis. Toward that end, state and federal health officials are trying to come up with better tools to identify outbreaks and to track down their sources more quickly. One prong of this effort is the recently expanded Food-borne Diseases Active Surveillance Network (FoodNet), a collaborative project of several federal agencies to establish an early-warning system for food-borne illness outbreaks. FoodNet consists of eight public laboratories scattered around the country from California to Connecticut, and government officials plan to add one or two more. These high-tech labs use pulsed-field gel electrophoresis ("DNA fingerprinting") and other sophisticated technologies to monitor the population for infections caused by *Salmonella, Shigella, Campylobacter, E. coli* O157:H7, *Listeria, Yersinia, Vibrio* bacteria species, and *Cyclospora* and *Cryptosporidium* parasites. Viruses will be added to the list in the near future, Drew Voetsch, CDC's FoodNet project coordinator, announced in May 1998. Computer networks that link and analyze the findings of these laboratories are designed to determine, for instance, whether victims of food-borne infections from different parts of the country were sickened by the same food source. Consumers and physicians can help make this system work by reporting suspected food-borne infections to their nearest public-health department.

In a related effort, the CDC is setting up PulseNet, a nationwide network of public-health laboratories that performs DNA fingerprinting on bacteria that may be food-borne. When public-health officials in one state identify a suspect bacterium, they will electronically send its DNA fingerprint pattern and related information to a central computer at the CDC. If laboratories in different locations submit the same DNA pattern during a defined time period, the CDC computer will alert PulseNet participants of a possible multistate outbreak so that a timely investigation can be done. According to Vice President Al Gore, PulseNet will help determine "as much as five times faster" if a certain food product should be recalled from supermarkets and restaurants. PulseNet is expected to be fully operational by 1999.

PulseNet's comparisons of DNA fingerprints will also quickly distinguish between a real outbreak and a "pseudo outbreak," or an increase in the number of cases that are not linked to a common source, CDC officials say.

In New Jersey, DNA fingerprinting was used toward that end in the summer of 1994 after state health officials had asked clinical laboratories to report all cases of *E. coli* O157:H7 to the health department. The number of reported *E. coli* O157:H7 infections increased nearly tenfold, and officials suspected an outbreak. Pulsed-field gel electrophoresis showed that the situation arose from increased reporting of cases and not from a common food source; seventeen different "fingerprints" were identified among bacteria isolated from twenty-three patients. According to the CDC, identifying pseudo outbreaks as such helps health departments conserve resources by investigating only true outbreaks.

Scientists, meanwhile, are racing to create more rapid laboratory tests to detect dangerous pathogens in food, and new discoveries are being made on a regular basis. One, which was recently announced by Springfield College in Massachusetts, is capable of detecting lethal strains of *E. coli* fast enough to enable processing plants to prevent distribution of contaminated food. According to the developers, the method identifies *E. coli* O157:H7 in meat in eight to twenty-four hours and may be adapted to identify *Salmonella*. The test kit is reportedly inexpensive, and using it does not require sophisticated equipment or advanced training. Barring patent or manufacturing problems, the kit is expected to become commercially available by the end of 1998. Another rapid food-safety test has been developed by Neogen Corporation to detect histamine, the intoxicant that causes scombroid fish poisoning. The test, called Veratox, produces results in one hour compared to up to eight hours for the conventional test, according to Neogen's LeeAnn Applewhite. Only time will tell whether these or other new rapid-testing methods meet expectations and prove to be useful additions to the existing food-testing technologies.

Of course, no laboratory test yields accurate results 100 percent of the time, the vast majority of foodborne illnesses will continue to go unreported, food production and distribution methods will continue to change, and new food-borne microbes will continue to emerge. Over the last two decades, more than thirteen disease-causing bacteria and viruses have been newly recognized as predominantly food-borne in the United States. As Robert V. Tauxe, M.D., M.P.H., an official with the Centers for Disease Control and Prevention, writes in the agency's journal *Emerging Infectious Diseases,* "Preventing foodborne disease is a multifaceted process without simple or universal solutions."

"If history has taught us anything," says Robert Buchanan, "every time we think we've got everything under control, something new pops up." He points to a naive assumption fifty years ago that the advent of antibiotics would mark the end of infectious diseases. Indeed, many scientists now theorize that the misuse of antibiotics in animals has led to the rise of some of the most dangerous and difficult-to-treat food-borne bacteria, such as multi-drug-resistant *Salmonella typhimurium* DT104, and a fluoroquinolone-resistant strain of *Campylobacter.* In a recent study published in the *New England Journal of Medicine,* CDC researchers note that the prevalence of *S. typhimurium* DT104, which is resistant to five antibiotics, has risen from 0.6 percent in 1979 and 1980 to 34 percent in 1996. "Multidrug-resistant *typhimurium* DT104 has become a widespread pathogen in the United States," the authors write.

"We're already seeing signs that other bacteria are becoming resistant to various forms of antibiotics," says Caroline Smith DeWaal of the Center for Science in the Public Interest (CSPI). Treating animals with low doses, or subtherapeutic amounts, of antibiotics benefits farmers because their animals grow faster and get sick less often, which means they can be crowded closer together. Indirectly, subtherapeutic antibiotic treatment may help keep meat and poultry prices in check by maintaining a more plentiful supply of these products. The practice also benefits the pharmaceutical industry, which has found a vast veterinary market for their antibiotic drugs. While acknowledging that new foodborne pathogens can also evolve naturally in the absence of antibiotics, CSPI nonetheless has repeatedly called on the government to ban subtherapeutic antibiotic use on the farm. Authors of the *S. typhimurium* DT104 study call for "more prudent use" of antimicrobial agents in farm animals and more effective disease prevention on farms so as to slow the dissemination of *S. typhimurium* DT104, to reduce the chances of this bug developing resistance to even more antibiotics, and to slow the emergence of more drug-resistant *Salmonella* strains. DeWaal expresses pessimism that the situation will change anytime soon, although CDC is clearly becoming increasingly concerned.

Worrisome Trends

The emergence of drug-resistant germs in the food supply is one of the two most worrisome trends that will dominate the food-safety arena for several years to come. The other trend is that familiar pathogens are showing up on an ever-widening array of food products. The most obvious example of this is *E. coli* O157:H7, which was previously thought to be found only in meat but is now menacing fresh produce and raw juices.

The good news is that food-borne illness surveillance techniques are improving all the time. For example, the sophisticated DNA "fingerprinting" technology that triggered the rapid recall of 25 million pounds of Hudson ground beef in 1997 would not have been available two years earlier. But public-health officials are not resting on their laurels, according to Buchanan. "Ideally, we would respond proactively—go through some kind of strategic planning process, asking where is the next problem likely to emerge," he explains. "That doesn't always work, but at least it gets people thinking about the fact that the bugs, the foods that can harbor these bugs, and the food-production technologies that we have today are not necessarily what we're going to have tomorrow."

New Production Technologies

Some new production technologies hold promise for making America's food supply safer. The technology that seems to offer the greatest potential is "competitive exclusion"—putting a group of harmless inhibitory bacteria into a farm animal's gut to prevent the growth of bacteria that may not make the animal sick but can cause disease in humans. In effect, the harmless bacteria occupy so much space and absorb so many nutrients that pathogenic bacteria cannot get a foothold.

PREEMPT, a blend of twenty-nine inhibitory bacteria, has been shown to drastically reduce the amount of pathogenic *Salmonella* in chickens when it is sprayed onto chicks and the chicks then ingest the bacterial cocktail as they preen. Approved in 1998, PREEMPT is too new a product to gauge its success in widespread use on chicken farms, but researchers are already trying to determine whether PREEMPT would work on other animals. Scientists are also formulating new potions of these "competitive microflora" to fight other pathogens in animals and on produce. Researchers at the University of Georgia recently reported that they could reduce or prevent the colonization of *E. coli* O157:H7 in cattle by treating the animals with eighteen harmless strains of *E. coli* when they are calves. The only problem with this approach is that, like an *E. coli* O157:H7 vaccine, it does not address the other Shiga-toxin-producing *E. coli* strains.

Other scientists, including a research team at North Carolina State University, are exploring the use of harmless lactic acid bacteria (LAB) as competitive microflora in various food products, including the surfaces of fruits and vegetables. LAB occur naturally on many foods, including yogurt and fermented vegetables. According to senior researcher Fred Breidt, when used as a preservative, LAB can prevent the growth of other bacteria without changing the food's taste, smell, or texture.

"It's widely known that competitive exclusion works," says Barry Swanson, a food-safety expert at the University of Washington. "But it's not known whether it works in commercial situations."

It is also widely known that irradiation is an effective way to sanitize food, but large-scale implementation continues to be bogged down in controversy. "We're not big fans of irradiation," says CSPI's DeWaal, suggesting that producers and processors need to attack microbial contamination at its source. "We don't think (irradiation) is the first place we should go to address food-safety problems. But if it's the only solution to *E. coli* O157:H7, then the industry should use it."

According to *Safe Food,* a book published in 1991 by CSPI, the most serious problems surrounding irradiation may have more to do with occupational and environmental risks than with food safety. "Irradiation requires more transport of radioactive materials on busy roadways, puts more workers at risk of exposure to low-level radiation, and adds to the hazardous-waste-disposal problem," the book says.

Some critics say irradiation may affect the flavor and texture of food, and that it can reduce its nutritional value. But supporters cite scientific studies demonstrating that irradiation does not change food's flavor or texture or reduce its nutritional content in any significant way. Furthermore, they say, when it comes to eliminating biological hazards in food, irradiation's value is beyond dispute. Michael T. Osterholm, state epidemiologist for the Minnesota Department of Health, is one of irradiation's staunchest advocates. "Just as thermal pasteurization of milk protected us from *E. coli* O157:H7 before we knew it was in raw milk," he writes in *Emerging Infectious Diseases,* "irradiation pasteurization can protect us from tomorrow's emerging food-borne pathogen."

Foods that have been irradiated must be labeled as such and must carry a logo that resembles a flower surrounded by a circle whose arc is broken across the top, according to FDA regulations. However, the meat industry has been lobbying to remove the labeling requirement, and the labeling issue has become a battleground between consumer groups and the food industry. The meat industry is reportedly reluctant to invest in irradiation technology unless it is convinced that the public will buy irradiated products. A 1997 industry poll showed that one third of U.S. consumers were reluctant to buy irradiated food, according to an industry newsletter, *Meat Industry Insights*. However, according to a November 1997 article in *Scientific American,* these days irradiation looks like the one relatively inexpensive technology that will allow beef packers to assure the safety of their products—"and it appears that a public now more fearful of tainted food than radiation may accept it."

A less controversial technology that found its way from the laboratory to industry is steam pasteurization. By briefly exposing animal carcasses to very high-temperature steam, it is possible to pasteurize the surface of muscle meat without cooking the product. The FDA's Buchanan says that some meat processors are already using steam pasteurization routinely, and there are a couple of experimental units that can pasteurize the surface of chicken carcasses in less than half a second.

If the meat or chicken is processed into a stew, it could be pasteurized through resistance heating, also known as ohmic heating. Ohmic heating is well suited for stews and other particulate food products, such as baked beans and prepared pasta dishes, because it uses an electrical probe to heat the product.

One of the problems with thermal pasteurization is that the heat may change the food's taste, appearance, or both. So engineers and scientists working in the food industry and in university settings are coming up with novel nonthermal pasteurization tools designed to kill pathogens and spoilage bacteria while maintaining the food's fresh flavor and appearance. Given Americans' penchant for fresh-tasting, minimally processed foods, the financial incentives to succeed have never been greater. Here are some of the new technologies being tested in laboratories or being tried commerically on a limited basis:

- **Bacteriocins.** Bacteriocins—proteins produced by safe bacteria—can inhibit the growth of pathogens, as well as of spoilage bacteria. Scientists have isolated bacteriocins for a food-sanitation approach similar to competitive-exclusion technology. One such bacteriocin is called Bifidocin B. It is produced by *Bifidobacterium bifidum*, a type of bacterium that inhabits the gastrointestinal tracts of breast-fed infants and healthy adults. According to *The Food Safety Consortium Newsletter*, jointly published by the Universities of Arkansas and Iowa, and Kansas State University, Bifidocin B and some other bacteriocins inhibit the growth of *Listeria monocytogenes* and two organisms that spoil cooked food. University of Arkansas food science Professor Michael Johnson says in the newsletter that the use of bacteriocins has become an increasingly attractive way to fight food-borne disease "partly because of consumer resistance to highly processed foods."

- **Pulsed electric fields.** Here, a food product is placed between two electrodes that pass different electrical charges between them. The electrical currents destroy bacteria by causing the cell membrane to deteriorate. One drawback is that if the bacteria contained a toxin, the toxin would be released. According to food scientist Barry Swanson, the only current commercial application for pulsed electric fields is liquid eggs. In the near future, pulsed electric fields may be used to pasteurize orange juice and other fruit juices.
- **Ozone.** If you have heard about the thinning ozone layer in the Earth's stratosphere, you may know that ozone is a highly reactive gas; its molecules are easily altered when mingled with certain air pollutants, such as chlorofluorocarbons. The same reactive properties that make ozone vulnerable in the stratosphere can also inactivate pathogens on food. It does this by weakening the bacterium's membranes and inactivating its metabolic systems. According to Swanson, the food industry is beginning to use low levels of ozone gas to reduce surface bacteria on poultry, meats, and fresh fruits. California Polytechnic University food scientists in San Luis Obispo, California, recently found that one ozone device consistently removed more than 99.99 percent of such pathogens as *E. coli* O157:H7, *Salmonella,*

and *Campylobacter* on surfaces of lettuce, meat, and poultry. The device's manufacturer, Tru-Pure Ozone Technologies of Yreka, California, makes commerical ozone units and a smaller version designed for consumers to use at home. The company's Internet address is http://www.longmarkozone.com/.

- **Oscillating magnetic field pulses.** This experimental method is designed to inactivate microorganisms on food after it is packaged in plastic or glass. The system exposes the product to a magnetic field that rapidly alternates, or oscillates, between positive and negative poles. This high-intensity oscillation seems to confuse the bacteria, interfering with their metabolism to the point that they cannot grow or reproduce. If validated scientifically, oscillating magnetic field pulses could potentially be used to sanitize almost any packaged food product.

In many instances, the best approach to making food safer is using multiple techniques. For example, the oxygen in a bagged salad can be replaced with nitrogen or carbon dioxide to retard spoilage. The manufacturer could also add some bacteriocins or competitive bacteria to inhibit the growth of pathogenic bacteria on the salad. Additionally, the salad could be treated with citric acid to lower its pH, making the product less conducive to bacterial growth. Of course, any and all food-safety technologies are most effective when coupled with safe food handling and good manufacturing practices.

Some low-tech, practical steps taken on the farm might also make a difference. Studies by Dale Hancock of the College of Veterinary Medicine at Washington State University suggest that *E. coli* O157:H7 survives and multiplies in wet feed and water troughs. This growth can potentially be slowed or halted, he says, by using mixed rations containing animal feed with high levels of certain acids, and by frequent cleaning and appropriate sanitation of water troughs.

The Price of Perfection

But what if the entire food supply could be irradiated or otherwise sanitized before reaching the consumer? Would that benefit public health in the long run? The answer is not as clear as it might seem. The upside of being periodically exposed to very low levels of food-borne pathogens is that we develop a natural immunity to a variety of bugs. This immunity may protect us if we accidentally cross-contaminate something in our kitchens, and it may reduce our risk of getting sick while traveling. It may seem farfetched, but if we had absolutely no immunity to, say, *Campylobacter jejuni* or *Salmonella enteritidis,* a bioterrorist could use one or both of these bacteria to spike a public water system or food supply. One of the critical questions scientists still need to address is whether people who consume irradiated food exclusively fail to develop a natural immunity to common food-borne pathogens.

Humans evolved to be omnivores; our digestive and immune systems are equipped to protect us against many insults. In fact, the human digestive tract is more akin to a pig's than to that of any other animal species. For the estimated 80 percent of Americans who are not at high risk for food-borne infections, using safe food-handling practices is probably all they need to avoid serious problems.

On the other hand, sterilized meat, poultry, seafood, and produce ought to be available to those who need it. After all, consumerism in America is all about choices. Grocery stores devote entire aisles to breakfast cereals; we can choose from sixteen varieties of spaghetti sauce and a hundred different types of frozen meals. Buying turkey wings, chicken breasts, pork chops, ground-beef patties, strawberries, and cantaloupes that have been irradiated or subjected to some other kill step ought to be one more choice for consumers. As a healthy adult, you may not want to spend an extra three or five cents a pound for sanitized food. But you might not mind if you are taking antibiotics or antacids, or if you want to do something extra to protect your elderly parent, young child, or diabetic husband.

Consumer demand for sanitized food options may grow as more of us slip into high-risk categories. The number of elderly people in this country is expected to double by 2030. There will be more organ recipients and more people undergoing cancer chemotherapy. Bone-marrow transplants will be used for a widening scope of diseases, and barring the discovery of a cure or vaccine, the AIDS population will undoubtedly increase.

At the same time, many food-safety experts believe

that the President's Food Safety Initiative, HACCP, increased consumer awareness, and other factors will ultimately bring down the rate of food-borne illnesses. But this won't happen overnight, and any reduction in human disease will be hard to prove. Michael Osterholm of the Minnesota Department of Health says that measuring the impact of HACCP and the President's Food-Safety Initiative on public health will be difficult, if not impossible "since we don't know what the numbers [of food-borne illnesses] are to begin with."

The problem may even appear to get worse before it gets better. Improved surveillance techniques may result in an increase in reported and confirmed outbreaks over the next four to five years. After that, the reported number of outbreaks and sporadic food-borne illnesses should come down as a result of stronger regulatory controls and public education efforts, predicts food scientist Douglas Powell, Ph.D., of the University of Guelph in Ontario, Canada.

Powell edits a daily electronic digest of news stories, press releases, and scientific papers relating to food safety. At last count, FSNet had some 3,000 subscribers who receive the digest free of charge by E-mail. Instructions on subscribing to FSNet and other electronic food-safety news services appear in Appendix A.

As FSNet subscribers and anyone who reads the newspaper know, a wealth of new food-safety information comes out every day. On the political front, things can change at any moment. The next Administration may not put as much emphasis on food safety as the Clinton Administration has. Federal food-safety budgets can shrink.

It is also possible that the government will pay even more attention to food safety in the future. One bill, called The Consumer Food Safety Act of 1998, would create a separate federal agency to do nothing but regulate the food industry. Among groups supporting the move is the Center for Science in the Public Interest. "The FDA has a huge mandate; it is responsible for approving all drugs, all medical devices and biologics, and now it's in charge of tobacco," says CSPI's DeWaal. "Food is somewhat different from the other things the FDA regulates because food plants require frequent inspections." There are about 53,000 food-processing plants in this country, she says, but the FDA and state health departments inspect only about 10,000 a year. "Food plants are not even required to register with the FDA," she continues, "so the FDA, from year to year, doesn't even know who they're supposed to inspect." CPSI is proposing that the Center for Food Safety and Applied Nutrition, the Food Safety Inspection Service, and the FDA's pesticide and seafood safety programs be rolled into the single agency. Congress has directed the National Academy of Sciences to investigate the feasibility of creating a single food-regulatory agency.

Constituents with strong opinions about such issues can contact their senators and representatives. To make this easier, CSPI's Web site—http://www.cspi.net.org—recently added an "activist page" that allows you to send a form letter devised by CSPI, or a personal letter, to members of Congress via E-mail.

Food-Safety Education

Ken Moore, executive director of the Interstate Shellfish Sanitation Conference, says the bottom line in making our food supply safer is not HACCP or any other form of government regulation. The bottom line, he says, is educating consumers.

"There are certain people who should not eat certain foods, and we need to find ways to educate these people so they are aware of their risk and can take precautions," says Moore, who manages South Carolina's Shellfish Sanitation Program.

A country's food-safety system is only as good as its weakest link. Therefore all links in the food-safety chain ought to be strengthened continually. Scientists must continue to learn new ways to identify and combat food-borne pathogens. Farmers need to improve their understanding of how these microbes can infect plants and animals on the farm. The food industry must keep abreast of the expanding body of scientific research affecting food production and processing. The trucking industry must work harder to monitor and maintain safe food temperatures during transit and to sanitize vehicles between hauls of foodstuffs. Food retailers must teach new employees how to prevent contamination in the store. The food-service industry must figure out ways to educate more food handlers whose actions or inactions can make customers sick.

Another group that needs more food-safety education is doctors, who often do not recognize food-borne infections in their patients. Says Powell: "Many doctors simply pat patients on the head and send them home" rather than ordering the appropriate tests or counseling high-risk patients to avoid certain foods.

While surveys show that food-safety awareness among American adults has never been higher, public education must continue, even when the current media hype over food-borne disease wanes. Teaching children to prevent food-borne infections is, as DeWaal puts it, "the next frontier in consumer education." One tack parents and teachers can take is to integrate food-safety issues into a generalized discussion or lesson plan about germs. When you remind your four-year-old to cover his mouth when he coughs so as to avoid spreading germs, or to wash his hands after using the bathroom, you can mention that uncooked chicken and meat, and unwashed fruits and vegetables, can carry germs, too. When Jill refused to buy her son, Jason, buffalo wings in the supermarket that day, she linked it to a conversation they had just had in the car on the way to the store. "I had a big frozen turkey defrosting in the refrigerator, and Jason said to put it on the counter so it would defrost faster," Jill says. "That launched me into this whole discussion on how parts of the turkey might warm up faster than others at room temperature, and how that could allow germs to grow on it. Then I explained that germs can also grow on cooked buffalo wings if they aren't kept hot enough."

Ideally, every school district would teach students about food safety. "Young people should be reached through age-specific school curricula, such as personal hygiene and special 'living skills' units that address food safety and diet," Christine M. Bruhn, of the Center for Consumer Research, writes in the CDC journal

Emerging Infectious Diseases. Whether her suggestion will be universally implemented remains to be seen. Every once in a while, food science Professor Donn Ward will chat with his wife, a fifth-grade teacher, about the importance of teaching kids about food safety. "She'll just throw her arms up and talk about how full the curriculum is already, teaching human growth and development, ethics, and so on," says Ward. "She asks, 'Where are we going to find the time to teach reading, writing, and arithmetic, much less food safety?' "

Educators who can carve out time for food-safety education have a variety of resources to turn to, including:

- The Food-borne Illness Educational Materials Database, a compilation of consumer and food-worker educational materials developed by universities; private industry; and local, state, and federal agencies. Among other things, the database includes computer software, audiovisuals, posters, games, and teaching guides for elementary and secondary school education. It can be accessed via the Internet: http://www.nal.usda.gov/fnic/foodborne/wais.shtml; or through the USDA/FDA Food-borne Illness Education Information Center, National Agricultural Library/USDA, Beltsville, MD 20705-2351; (301) 504-5719/Fax (301) 504-6409; E-mail: fnic@nal.usda.gov.
- *Don't Get Bugged by a Food-borne Illness,* a game developed by the University of Nebraska Cooperative Extension. Each game packet

contains a "quiz bowl" version to use with single players at health fairs, school fairs, or clinics, plus a "bingo" version to use with groups in any setting. All materials may be reproduced on a standard copier for unlimited use. Designed to be used by adults with youths over age twelve, *Don't Get Bugged* has received an award from the National Extension Association of Family and Consumer Sciences. According to Extension educator Alice Henneman, more than 1,300 teachers nationwide have used *Don't Get Bugged* since it was introduced in September 1996. To order the game, send $15.95 ($13.95 plus $2 for shipping) to: University of Nebraska Cooperative Extension in Lancaster County, 444 Cherrycreek Road, Lincoln, NE 68528-1507. If you have questions about ordering, call (402) 441-7180.

- Food-safety courses on the Internet. One is titled "Safe Food: It's Up to You!" and can be found at http://www.extnet.iastate.edu/Pages/families/fs/Lesson/Lessonfs.html. Designed by the Iowa State University Extension Office, the course has been completed by more than 3,000 people worldwide and is being field tested in fifteen Iowa high schools. The four-lesson course features a comic strip set in the "Artichoker's Café," where the owner learns about the dangers of food-borne illness as he prepares a Thanksgiving turkey.
- Fight BAC! The Partnership for Food Safety

Education has a variety of educational objectives, one of which is the development of a food-safety curriculum for middle schools, according to Robert Curry, Fight BAC! spokesman. An unveiling of the new curriculum is anticipated for early 1999. For information, call (202) 429-8273 or E-mail fightbac@mindspring.com. The Fight BAC! Web site has a wealth of information for consumers and educators, and it is updated on a regular basis. The Web address: http://www.fightbac.org.

- An on-line coloring book for children focusing on food safety at home, school, and when eating out. This publication, which includes several activity pages, a letter to parents, and a certificate of participation, is cosponsored by the FDA, USDA, and the Chef and Child Foundation. The Web address is: http://www.foodsafety.gov/~dms/cbook.html.
- *Germ Squirm—Kids and Safe Food Handling,* from the Virginia Cooperative Extension. This four-page guide for teachers is divided into activities and questions for two- and three-year-olds, three- to five-year-olds, and four- to six-year-olds. Among other things, activities teach children to identify dirt and places where germs live and hide; to identify good and bad germs and learn how to get rid of germs; and to classify ways to fight germs while eating out. It also demonstrates steps for washing hands, utensils, and food, and safe food-storage

techniques. To obtain a copy, send $1 to Extension Distribution Center, 112 Landsdowne Street, Blacksburg, VA 24061-0512. For more information, call (703) 231-4673.

- "Discovering Food Safety—Detective Mike Robe's Fantastic Journey for Head Start Preschoolers." This curriculum, developed by the University of Rhode Island Cooperative Extension Service, includes a teaching outline for six lessons, student activities, and a ten-minute videotape. Cost is $65. For information, contact the University of Rhode Island, Department of Food Science and Nutrition, 21 Woodward Hall, Kingston, RI 02881; (401) 874-2972.
- "Adventures with Mighty Egg," an integrated curriculum unit for grades K–3 featuring a hands-on approach designed to encourage students to want to know more about eggs and other subjects as they develop math, science, language arts, creativity, and other skills. Safe handling of eggs is integrated in some of the lessons. Curriculum includes stickers, poster, activities, recipes, eight lesson plans, and a reading list. Cost is $9 per kit (less if ordered in quantity); extra posters and stickers are $3.50. Order kit No. 276 from the American Egg Board, 1460 Renaissance Drive, Park Ridge, IL 60068; (847) 296-7043/Fax: (847) 296-7007; E-mail: aeb@aeb.org.
- "Glitter Bug," a kit that teaches proper hand-washing techniques to schoolchildren. Materials

include powder or potion, ultraviolet light, manual, and motivation cards. Students put powder or potion on hands, wash, then put hands under UV light. Any powder or potion remaining on the hands glows. Teaching materials then show proper hand-washing techniques. Cost varies from $65 to $245. For a brochure, contact Brevis, 3310 South 2700 East, Salt Lake City, UT 84109; (800) 383-3377/Fax: (801) 485-2844; E-mail: brbrevis@xmission.com.

- *Play It Safe: Goals for Food Safety,* a sixteen-page workbook with activities, game board, four lesson plans, and suggested ways to use materials with appropriate textbooks. Cost is nominal. Contact Publications Sales Department, Food Marketing Institute, 800 Connecticut Avenue, NW, Suite 500; Washington, DC 20006-2701; (202) 429-8298/Fax: (202) 429-4529; Web site: http://www.fmi.org/. (Publication may be out of print, but can still be borrowed from the National Agricultural Library.)
- The EPA's Office of Ground Water and Drinking Water has compiled a list of on-line links to resources for kids interested in learning about drinking water. The list, which also contains links for teachers, can be accessed at: http://www.epa.gov/OGWDW/kids/index.html.

Another way to encourage food-safety education is to ask your child's school to designate a Food-Safety Awareness Day or Food-Safety Awareness Week. County extension offices, land- or sea-grant universities, and food-processing companies may be able to provide speakers or lesson plans.

If children learn and practice good food-safety habits from a young age, they are more likely to carry those habits into their adult lives—and to teach them to their own children.

GLOSSARY

adulterant: bacterium or other contaminant that the federal government considers illegal in food

assay: laboratory test or analysis

bacteremia: blood poisoning caused by bacteria in the bloodstream

bacteria (singular: bacterium): living single-celled organisms that can be carried by water, wind, insects, plants, animals, people

bismuth subsalicylate: the active ingredient in Pepto-Bismol that treats gastrointestinal symptoms

bivalves: marine animals with two shells hinged at one end (clams, oysters, mussels); also known as mollusks or molluscan shellfish

chlorination: method of disinfecting water by adding chlorine gas or solid hypochlorite

colonization: the proliferation of bacteria in the gut

colony-forming unit: single live bacterium capable of growing and dividing

competitive exclusion: system that introduces enough harmless competitive bacteria into an animal to prevent the growth of bacteria that can cause human disease

complication: secondary symptoms or diseases (often worse than the primary symptoms) that result from an infection

contamination: the unintended presence of potentially harmful microorganisms in food

cross-contamination: transfer of disease-causing microorganisms by hands, cutting boards, countertops, sponges, cloth towels, aprons, or utensils. Cross-contamination can also occur when juices from raw animal products drip onto cooked or ready-to-eat foods

cytokines: molecules produced by the body in response to the invasion of disease-causing microorganisms

diarrhea: loose or watery bowel movements that occur with increased frequency and increased volume; typically caused by a food-borne infection

disinfection: treating water (usually with chlorine or ozone) to inactivate, destroy, or remove disease-causing bacteria, viruses, and protozoa or other microscopic parasites

disinfection by-products (DBPs): potentially hazardous chemicals that are created in water when disinfectants, such as chlorine, react with organic material, such as leaves or soil

enhanced coagulation: a process that combines particles in water to help remove organic matter

fecal-oral route: getting a microscopic amount of infected fecal matter on the hands and then transferring the infection to the digestive tract by putting the hands in the mouth

filtration: water-treatment step used to remove small particles, dissolved organic material, odor, taste, and color

food-borne illness: infectious disease transmitted to humans by harmful substances in food. Examples: salmonellosis caused by *Salmonella* bacteria, and botulism caused by the toxin produced by the bacterium *Clostridium botulinum*

food-contact surface: any equipment or utensil that normally comes in contact with food or that may drain, drip, or splash onto food or onto surfaces normally in contact with food. Examples: cutting boards, knives, sponges, countertops, and colanders

food poisoning: a generic term for food-borne infections; technically refers only to illness that occurs after ingesting a preformed toxin

fungi: group of microorganisms that includes molds and yeasts

gastroenteritis: an inflammation of the stomach or intestinal lining that usually results in sudden, violent episodes of diarrhea, vomiting, or both; a common symptom of food-borne infection; also called GI distress

Guillain-Barré syndrome: neurological complication of certain food-borne infections that causes peripheral nerve damage, numbness, and paralysis

HACCP (pronounced "HASS-ip"): acronym for Hazard Analysis and Critical Control Points, a scientifically based, federally mandated food-safety system used by most segments of the food industry

incidence: the number of new cases of food-borne illness in a given population during a specified period of time

incubation period: amount of time it takes after exposure to a germ for symptoms to emerge

infectious dose: number of bacteria, viruses, or protozoa required to cause disease

intoxicant: chemical in food that causes food poisoning; example: histamine

irradiation: exposing a food product to low-level radiation in order to kill or inactivate microorganisms; also called radiation pasteurization

kill step: any processing method that destroys germs and spoilage bacteria in food. Examples: irradiation, pasteurization, boiling, or cooking.

microbial contamination: situation that occurs when food is tainted by disease-causing bacteria, viruses, parasites, or fungi, or their toxins; certain plants and fish produce their own natural toxins

micrometer: one-millionth of a meter; sometimes called micron

microorganism: life-form too small to be seen without a microscope. Examples: bacteria, fungi, protozoa, viruses

oocyst: egg of a protozoan

outbreak: incident in which two or more people experience the same illness due to the same microorganism after eating the same food

parasite: microorganism that needs a living host in order to reproduce; examples: *Cryptosporidium, Toxoplasma* species

pasteurization: a process to kill pathogens in food, usually by heat

pathogen: any microorganism that is infectious and causes disease; informally known as a germ or bug

peptic acid: digestive acid that breaks down food and may destroy disease-causing bacteria

pH level: a measure of acidity; the lower the pH, the more acidic the food or beverage

preformed toxin: poison manufactured by a bacterium while it is still in the food

protozoan (plural: protozoa): single-celled animal

pulsed-field gel electrophoresis: laboratory technique that generates a DNA "fingerprint"

radiation pasteurization: see *irradiation*

reactive arthritis: painful joint inflammation that can set in after the initial symptoms of certain food-borne infections, such as salmonellosis

residual disinfection: a required level of disinfectant that remains in treated water to ensure disinfection protection and prevent recontamination throughout the water-distribution system

septic shock: life-threatening condition caused by tissue damage and a dramatic drop in blood pressure as a result of bacteria and their toxins in the blood

serotype: group of very closely related microorganisms

Shiga toxin: a poison produced in the body by certain types of infectious bacteria including *E. coli* O157:H7

spoilage bacteria: microorganisms that make food look, smell, and feel rotten but generally do not cause human disease

spore: thick-walled protective structure produced by certain bacteria and fungi to protect their cells. Spores often survive cooking, freezing, and some sanitizing measures

temperature abuse: failing to keep hot foods hot enough or cold foods cold enough to kill pathogens or inhibit their growth; common cause of food-borne illness

temperature "danger zone": 40°F to 140°F—the range of temperature at which most disease-causing bacteria can grow and multiply, although some species, such as *Listeria,* can grow at refrigerator temperatures

toxin: poison produced by certain species of microorganisms, fish, or plants; example: botulin toxin from *Clostridium botulinum*

trihalomethanes (THMs): most common class of disinfection by-products; created when chlorine reacts with organic matter in water during the disinfection process

virulence: degree to which a microorganism can make a person sick

virulence factor: means by which a disease-causing microorganism causes symptoms

virus: protein-wrapped genetic material that is the smallest and simplest life-form known; examples of viruses that can be food-borne: Norwalk viruses, hepatitis

A. Unlike bacteria, which can grow and multiply on food, viruses replicate by invading cells of their human or animal host

water-borne disease: an illness caused by a bacterium, virus, protozoan, or other microorganism capable of being transmitted by water, examples: typhoid fever, cholera, cryptosporidium

APPENDIX A
RESOURCES

Safe Tables Our Priority (STOP)
335 Court Street, Suite 100
Brooklyn, NY 11231
(800) 350-STOP
E-mail: feedback@stop-usa.org
Web address: http://www.stop-usa.org

Nonprofit grassroots organization devoted to victim assistance, public education, and policy advocacy for safe food and public health. The organization was founded in 1993 by family and friends of people who became ill or died from exposure to E. coli *O157:H7 and other pathogenic bacteria in meat and poultry.*

Lois Joy Galler Foundation for Hemolytic Uremic Syndrome, Inc.
734 Walt Whitman Road
Melville, NY 11747
(516) 673-3017
E-mail: bob@loisjoygaller.org
Web address: http://www.loisjoygaller.org

Founded by parents of a three-year-old girl who died of HUS in 1992, the foundation provides support, news, literature, and networking to HUS victims, survivors, and their families. The organization also raises funds for HUS research. Newsletter available. HUS videotape available for $49.99 plus postage.

Center for Science in the Public Interest (CSPI)
1875 Connecticut Avenue, NW, Suite 300
Washington, DC 20009
Phone: (202) 332-9110, fax (202) 265-4954
E-mail: cspi@cspinet.org
Web address: http://www.cspinet.org

Nonprofit education and advocacy organization that focuses on improving the safety and nutritional quality of the food supply. CSPI seeks to promote health through educating the public about nutrition and alcohol; it represents citizens' interests before legislative, regulatory, and judicial bodies; and it works to ensure that advances in science are used for the public's good.

USDA Complete Guide to Home Canning
Available on-line at: http://www.foodsafety.org/canhome.htm, or as a 71-page spiral-bound booklet for $10 (Master Card or Visa orders only). To order, call (800) 226-1764 and request "Order No. SF06."

This guide provides complete information on canning numerous types of foods, and explains the principles behind canning. Included are easy-to-use tables of different processing times for homes located at various altitudes.

USDA Meat and Poultry Hotline, (800) 535-4555

Staffed year-round by home economists, registered dietitians, and food technologists from 10 A.M. to 4 P.M. eastern standard time, the hotline answers consumer questions about buying, storing, and cooking meat, poultry, and eggs. An extensive selection of food-safety recordings can be heard twenty-four hours a day using a touch-tone phone. Information and publications can be downloaded from USDA's Home Page on the Internet at http://www.usda.gov/fsis or by calling (202) 690-3754/5 or toll-free at (800) 238-8281.

FDA Seafood Hotline, (800) FDA-4010 or (202) 205-4314 in Washington, DC

Provides expert advice about buying, storing, and preparing seafood. The automated hotline and Flash Fax service are available twenty-four hours a day. Product recalls and alerts are played first. Spanish menu available. Public affairs specialists can be reached Monday through Friday from noon to 4 p.m. eastern standard time.

USDA/FDA Foodborne Illness Education Information Center

10301 Baltimore Boulevard, Room 304
Beltsville, MD 20705-2351
Phone: (301) 504-5719
E-mail Cindy Roberts, information specialist: croberts@nalusda.gov
Web address: http://www.nal.usda.gov/fnic/foodborne/foodborn.htm

Provides information about food-borne illness prevention to individuals and groups developing educational and training materials for food workers and consumers.

EPA Safe Drinking Water Hotline, (800) 426-4791

Provides explanations of drinking water standards and copies of health information relative to specific drinking water contaminants.

Institute of Food Technologists Web site

Web address: http://www.ift.org

E-mail: info@ift.org

Founded in 1939, the Institute of Food Technologists is a nonprofit scientific society with 28,000 members working in food science, food technology, and related professions in industry, academia, and government.

Iowa State University Extension Food Safety Project

Web address: http://www.exnet.iastate.edu/Pages/families/fs/

Provides news and educational material designed to help consumers minimize the risk of food-borne illness.

North Carolina State University Food-Safety Web site.

Web address: http://www.ces.ncsu.edu/depts/foodsci/agentinfo

FoodTalk Electronic Newsletter

Published by the University of Nebraska Cooperative Extension, this free monthly E-mail newsletter provides short "how-to" messages on food, nutrition, or food safety for health professionals, educators, and consumers.

To subscribe:

1) Send E-mail to listserv@unlvm.unl.edu
2) Leave the subject blank
3) Type message: SUBSCRIBE FOODTALK
4) Do not include signature when subscribing

FSNet

An electronic digest of articles related to food safety—including microbial hazards, nutritional issues, and regulatory issues—culled from journalistic and scientific sources around the world.

To subscribe:

1) Send E-mail to: listserv@listserv.uoguelph.ca
2) Leave subject line blank
3) In the body of the message, type: subscribe fsnet-L firstname lastname (i.e., fsnet-L Doug Powell)

Agnet and AnimalNet

Like FSNet, Agnet and AnimalNet are distributed daily by E-mail to thousands of individuals from academia, industry, government, the farm community, journalism, and the public at large. Material related to plant agriculture—food biotechnology, chemical hazards, productivity and sustainability—is included in Agnet. Material related to animal agriculture—including new diseases, sustainability, and animal welfare—is included in AnimalNet.

To receive Agnet or AnimalNet:

1) Send an E-mail message to:
   listserv@listserv.uoguelph.ca
2) Leave subject line blank
3) Type agnet or animalnet followed by L firstname lastname (i.e., subscribe agnet-L Doug Powell)

APPENDIX B
FURTHER READING

How to Prevent Food Poisoning: A Practical Guide to Safe Cooking, Eating, and Food Handling, by Elizabeth Scott and Paul Sockette (John Wiley & Sons, 1998)

Diet for a New America: How Your Food Choices Affect Your Health, Happiness, and the Future of Life on Earth, Second Edition, by John Robbins (H J Kramer, 1998).

Food Safety (*True Book*) by Joan Kalbacken (Children's Press, 1998) (reading level: ages 4–8).

Cooking for Life: A Guide to Nutrition and Food Safety for the HIV-Positive Community, by Robert H. Lehmann (Dell, 1997)

Spoiled: The Dangerous Truth About a Food Chain Gone Haywire, by Nicols Fox (Basic Books, 1997)

Mad Cow U.S.A.: Could the Nightmare Happen Here? by Sheldon Rampton and John C. Stauber (Common Courage Press, 1997)

Safe Food: Eating Wisely in a Risky World, by Michael F. Jacobson, Lisa Lefferts, and Anne Witte Garland (Living Planet Press, 1991)

INDEX

D

G

K

L

M

N

O